W1FB's Antenna Notebook

by Doug DeMaw, W1FB

Published by the
AMERICAN RADIO RELAY LEAGUE
225 Main Street
Newington, CT 06111

CONTENTS

Foreword

FOREWORD

I have prepared this text for amateurs rather than engineers or scientists. You will find the material easy to understand, even though you do not have a strong technical base from which to work.

We shall explore the many aspects of simple antennas and related matters. A large amount of this material is not found in other antenna books. The subjects I treat in this volume are important to beginners as well as experienced amateurs. You will learn how to build and adjust a variety of common antennas that can provide excellent results for local and DX operation. I have chosen the plain-language approach in order to avoid confusion for those of you who do not have a background in antenna design and application. These antennas are founded on practical examples that I have developed and used for my amateur work. Most of the designs were conceived before my time as an amateur, so I am passing along to you the short cuts to easy fabrication and effective use. Some of the systems I treat in this book are original designs that I have found useful for specific kinds of communication under a variety of propagation conditions.

Another objective I had while writing this text was to dispell some of the common misconceptions about antennas that are passed along over the air and at hamfests. Free advice is not always accurate, despite the good intentions of the Elmer who offers his assistance!

I have not included descriptions of exotic gain antennas. Rather, I have emphasized, for the most part, simple wire and tubing antennas that can provide satisfactory perform- ance for a host of operating objectives. Antennas of elaborate design may be found in **The ARRL Antenna Book.**

There is no high-level math in this publication. You will find simple equations where they are necessary to explain a concept or to calculate the length of an antenna element of its matching section. Explicit drawings with numerous labels are used in an effort to clarify the illustrations to the fullest measure.

I hope you will have as much fun reading this book as I had while preparing it. If you enjoy constructing antennas, this volume will be of considerable help to you.

Chapter 1

SOME FUNDAMENTAL ANTENNA DATA

There are a number of things that you should know before launching an antenna-construction program. First, and no doubt of greatest importance, we should dispell some common myths about antennas and how they perform. There is little point in erecting an antenna if it does not do what you have been told it will do. In this regard there are many popular misconceptions that we should rid ourselves of. This chapter is devoted to this subject, along with some other basic information you should be given.

Antenna Directivity

How many times have you heard someone say, "My signal would be louder at your QTH if my dipole was broadside to you." This belief is fundamentally correct if the half-wave dipole or similar wire antenna is high above ground. Generally, we are considering a height of 1/2 wavelength or greater. Therefore, if we are operating on 75 or 80 meters, there will be considerable broadside directivity (figure-8 type of pattern) if our dipole is 120 feet or greater above earth ground. Most 80-meter dipoles are from 30 to 60 feet above ground, owing to available support poles, towers or trees.

You may be wondering, in view of the above statements, what the dipole pattern will be at the lower heights. We may generalize by saying that the closer to the ground we place the antenna, the less directive it becomes. The radiation pattern becomes pretty omnidirectional, and the radiation angle is high. If we could see the RF energy being radiated, it would appear like a ball around the antenna. This is good for short-range communications out to, say, 400 or 500 miles.

Take for example the case of a 160-meter dipole that is erected only 30 feet above ground. A half wavelength for this band is roughly 250 feet (pretty high!). This situation is equivalent to placing your 10-meter dipole only 3 feet and 10 inches above ground! None of us would be willing to do that if we wanted to be heard at great distances! Although 40 or 50 feet seems high to us as we look upward, this height becomes less and less meaningful as the operating frequency is lowered.

The antenna height above ground also affects the feed imped-ance of our antenna. For a half-wavelength dipole it can vary from 10 to 100 ohms over a perfectly conductive ground. A dipole that is 1/2 wave above ground has a characteristic impedance of approximately 60 ohms. At 1/4 wavelength above ground the impedance is roughly 70 ohms, but it will be on the order of 25 ohms if the dipole is only 1/8 wavelength (31 feet at 3.9 MHz) above actual ground. Along with this condition we will have an antenna that is not directional in the classic sense.

The foregoing discussion relates to most horizontal antennas for the MF (medium frequency) and HF (high frequency) bands. Vertical antennas are not affected by these rules, but do perform best when they are high above ground with an above-ground radial system. The advantage of height in this case is that the antenna is clear of nearby conductive objects, such as power lines, metal fences and such, which can distort the radiation pattern and absorb RF energy. Vertical dipoles and 1/4-wave vertical antennas both help to overcome the problem of antenna height. They are worth considering when you are unable to erect a horizontal antenna at a proper height above ground.

Where is True Ground?

We should always be aware that the conductivity of the earth varies with the geographical location. The true earth ground may, therefore, exist a few inches below the surface of the earth, or it may be several feet into the earth. This helps to explain why some hams with low antennas report phenomenal results, whereas other amateurs have poor results with the same type of antenna at a similar height. Therefore, it is not uncommon to hear someone say, "Golly, Joe sure has a hot radio location."

Generally, wetland areas, ocean shores and swamps provide good ground conductivity near the surface of the earth. Conversely, true ground may be quite a distance down in locations that are arrid and sandy or rocky. The mineral content in the earth plays an important roll in this respect also.

Artificial Grounds

Some amateurs, in an effort to establish a known ground plane, use a system of radial wires under an antenna. They may call these wires "radials" or a "counterpoise." The counterpoise reduces the influence of true earth while reducing the ground currents. This aids the antenna efficiency by lowering the system losses.

We must remember that the use of a ground screen or counterpoise lowers the effective height of the antenna. In other words, the distance between the antenna and the counterpoise is the effective height above ground. Ground screens are normally elevated a few inches or feet above the surface of the earth and are held aloft by means of poles and insulators. Some amateurs bury the wires in the ground, or simply lay them on the ground. See Fig. 1-1.

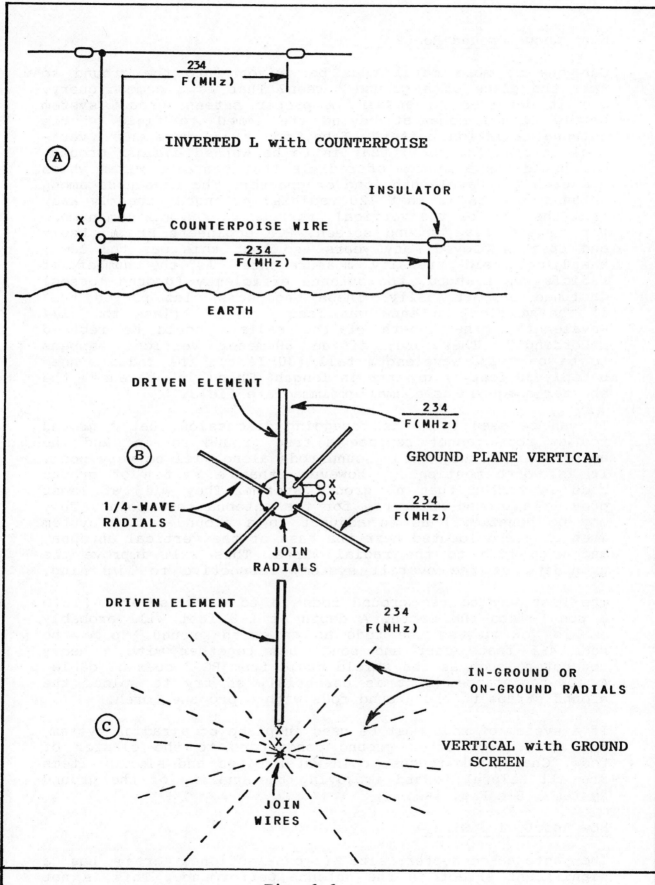

Fig 1-1

What About Ground Rods?

Can one or more metal rods be driven into the ground to take the place of a ground screen? That is a common query, and it deserves an answer. A proper antenna-ground system should extend somewhat beyond the immediate field of the antenna radiation pattern. Thus, for a 1/4-wavelength vertical antenna (Marconi) that is to be worked against ground, we should use a system of radials that contains wires which are each 1/4 wavelength long or greater. The consensus among antenna experts is that 120 radials, extended linearly away from the base of the vertical, represents the maximum number for an effective ground screen. RCA engineers Brown, Lewis and Epstein proved many years ago that this was the case, based on field-intensity measurements. As the number of radials was reduced, the antenna efficiency (ground losses) declined proportionally. These engineers also proved that if the vertical antenna was less than 90° (less than 1/4 wavelength), the length of the radials could be reduced accordingly. Therefore, if an 80-meter vertical happens to be only 1/8-wavelength tall (30 feet), the radials need be only 30 feet or greater in length. This is because the shorter antenna has a smaller immediate field.

It can be seen from the foregoing discussion that a ground rod or rods cannot replace a true ground screen, and the antenna efficiency with ground rods alone will be very poor. It is worth mentioning, however, that a system of ground rods is better than no ground system. They will at least provide a ground-reference for the antenna feed point. They may be beneficial as an adjunct to a ground-radial system when they are located near the base of the vertical antenna, and when tied to the radial wires. This will improve the grounding of the overall system, respective to lightning.

The best way to use ground rods is to drive several (4 to 8 rods) into the earth. A depth of 6-8 feet will probably enable you to use the rods as an earth ground. Space the rods 4-6 feet apart and bond them together with a heavy conductor, such as the shield braid from RG-8 coaxial cable. A low-resistance joint is essential, so try to solder the ground straps to the ground rods with a propane torch.

If a system of rods must be used in place of a radial system, try to connect other ground conductors to the cluster of rods. Chain-link fences, cold-water pipes and similar items are all helpful toward improving the quality of the ground system. See Fig. 1-2.

How About Wire Size?

The antenna-conductor size (cross-sectional area) has a significant effect on the antenna performance. This is not true of the conductors in a ground-radial system. The reason-

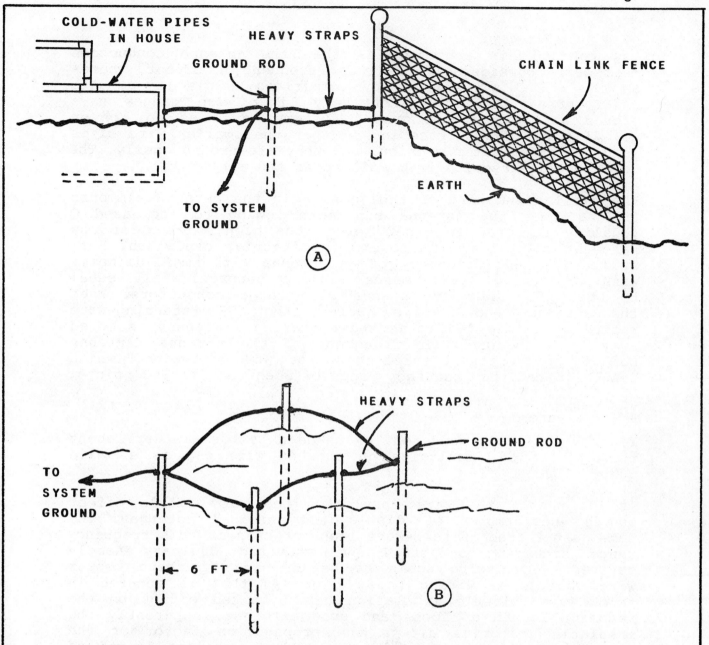

Fig. 1-2 -- A less-than-ideal ground may be formed by bonding as many conductors as possible (A). A poor ground is better than none, and good results are frequently had with a system like that at A. Illustration B shows how to use several ground rods to form an earth-ground reference for grounding the station equipment, or for use with an antenna that is worked against an earth ground. Better efficiency will result when using the ground system seen at C of Fig. 1-1. Chain-link fences and the cold water pipes in a home may be added to the ground system at B to improve it.

ing behind this statement is that the antenna conductors must carry considerable RF (radio frequency) current, whereas the radial wires have, by comparison, miniscule amounts of RF current flowing on them. Very light wire gauges have proved satisfactory in ground systems, such as no. 24 or no. 26 wire. The shortcoming is that the smaller wire sizes lead to faster deterioration, and they are broken easily. The deterioration results from soil acids and alkalinity.

The smaller the wire or tubing size in the antenna elements the narrower the antenna bandwidth, owing to increased Q (quality factor). In other words, the higher the Q of any resonant circuit the narrower its effective bandwidth. But, at the resonant frequency of an antenna with small-diameter wire or tubing, performance will be approximately equal to that of a similar antenna with large conductors. When the antenna bandwidth is narrow, the SWR (standing-wave ratio) will rise faster as the operating frequency is moved away from the resonant frequency of the antenna. Antenna bandwidth is often referred to as the "SWR bandwidth," which is considered the frequency range between the 2:1 SWR points.

<u>Insulated or Bare Wire?</u>

You have probably wondered, or asked a fellow amateur, about the effects of insulation on antenna wire. It is a valid query, indeed.

At MF and HF the wire insulation has very little effect on the performance of dipole and vertical antennas. The same is not true of full-wave loop antennas in this frequency range, according to tests I have conducted. Ordinary enamel-covered copper wire seems to be on par with bare copper wire, but house wiring (no. 12 or 14) that is encased in thick vinyl-plastic insulation has a marked effect on the resonant length of loops and quad antennas. Apparently the insulating material alters the propagation factor or the antenna element, as is the case with a resonant length of coaxial cable (0.66 velocity factor, approximately). This means that the wire must be shorter for a resonant condition than is the case when using bare wire. Apart from this phenomenon, the antenna performance seems to remain the same in either situation.

The propagation or velocity factor of heavily insulated wire becomes quite pronounced at VHF, UHF and higher. Hence, we must keep this in mind when designing antennas that are affected by the type of wire insulation used.

It seems to make no appreciable difference when using wire with or without insulation when constructing a ground system. Radial wires, for example, need not be resonant at the operating frequency.

Coaxial-Cable Considerations

We can get into a lot of trouble when working with the wrong type of coaxial feed line. The main consideration is line loss in dB (decibels), but other factors enter into the scheme of things when it comes to antenna-system efficiency and performance.

Not only is small coaxial cable, such as RG-58 and RG-59 lines, lossy in terms of dB, it can not accommodate high RF power when the SWR is much greater than 1.5:1. In the presence of a high SWR at, say, 300 watts, the standing waves on the feed line will cause hot spots that can actually melt the cable and cause it to bulge like a baloon! At points along the line where high RF voltage is present, the cable can break down from arcs between the inner and outer conductors. I have personally used RG-58 at 700 watts of transmitter output power, and no adverse effects were noted. But, this was in situations where the SWR was close to 1:1.

By way of an example, let's compare the attenuation in dB of 100-foot lengths of RG-174 (the smallest coaxial cable), RG-58A (medium diameter) and RG-8/U (large foam-filled line). The operating frequency in this example is 30 MHz. The loss in dB for RG-174 is 5.5. It is 2.5 dB for RG-58A and only 0.9 dB for RG-8/U. What does this really mean? Well, if you are operating at 10 meters with 100 feet of RG-174 cable, less than 50% of your transmitter power will reach the feed point of your antenna. Thus, if the transmitter output power is 50 watts, only 14 watts will arrive at the antenna! On the other hand, and under the same conditions, RG-8/U cable will allow 40.5 watts to reach the antenna. If we were to use the RG-58A cable we would find 28 watts of power at the antenna feed point.

The lowest losses in conventional coaxial cable are found in 50- and 75-ohm aluminum-jacketed "hardline." For example, we will find the attenuation for 75-ohm RG-247/U (100 feet) at 30 MHz a mere 0.25 dB, which yields 47 watts at the antenna when 50 watts are supplied to the transmitter end of the line.

As the operating frequency is lowered, the losses per 100 feet of line decrease. Conversely, the higher the operating frequency the greater the line loss. Open-wire feed line has the lowest loss of the many feeders available: 300-ohm open-wire line (1 inch spacing) has only 0.18 dB of attenuation at 30 MHz for 100 feet of line, but 300-ohm tubular feed line exhibits a loss of 0.45 dB at 30 MHz when the line is clean and dry.

Contaminated Coaxial Cable

Each of us has at some time gotten into trouble with old

or surplus coaxial or solid 300-ohm twin line. What are the symptoms of defective feed line? Unfortunately, there may be no outward symptoms that tell us we have a feed-line problem.

The end result of bad feed line is high loss. This comes as the result of the dielectric material (insulation) becoming contaminated by exposure to the sun (UV rays) over a long period, and exposure to pollutants in the air. These agents destroy the insulating materials within the cable, and this results in high power loss. The polyvinyl outer jacket of the line is affected by chemicals and UV rays, and the poisoned jacket causes a migration of chemical agents into the inner polyethylene insulation, thereby causing it to become a poor insulator.

Coaxial cables that are buried in the ground are especially prone to contamination that is caused by the acidity and alkalinity of the soil. The greater the soil moisture the faster the contamination process. Cables may become quite lossy in a matter of months, or it might take a couple of years for this problem to develop. It depends upon the condition of the soil in a particular region.

You can test your cable for loss by placing an RF wattmeter and a dummy antenna at the far end of the cable. Another RF wattmeter is used at the input end of the cable under test (see Fig 1-3). A known amount of RF power is fed into the line and the power at the far end is measured by the second wattmeter. Deduct the normal line loss in dB versus the operating frequency (dB converted to watts) from the power reading at the dummy antenna. If the overall loss in watts is greater than the expected line loss per 100 feet, you may consider the cable to be contaminated. When you perform this test, be sure to terminate the cable with a dummy antenna that has the characteristic impedance of the line (50-ohm load for 50-ohm cable, etc.).

The Importance of Line Loss

You should not underestimate the importance of low-loss cable in your antenna system. Cable losses absorb power and that means less of your signal is radiated from your antenna. It also means a signal loss in your receiver.

The situation is even worse if you have an impedance mismatch between the feeder and your antenna. Greater line losses mean greater total loss as SWR increases. Here are some examples of this phenomenon.

Consider a case where you are going to operate on 10 meters. You will be using a bit over 100 feet of RG-58 coax cable. Coax line losses are 3 dB when operating into a matched (50-ohm) load. That means that half your transmitter power is lost in the coax cable as heat, and your signals are about half a S unit weaker than they would be if there

9

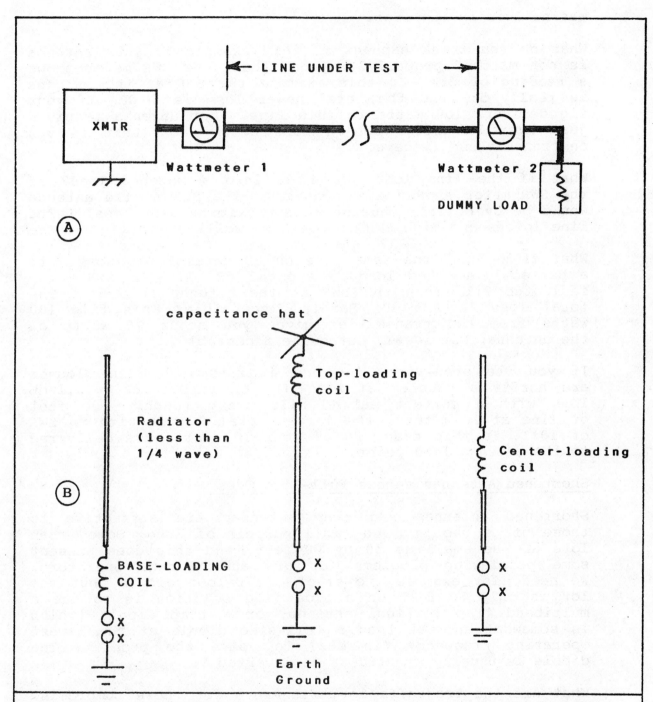

Fig 1-3 -- Illustration A shows how to test old coaxial line for losses caused by chemical contamination (see text). Shortened antennas (B) can be resonated by means of base, top or center loading. The coils seen above accomplish this. They are adjusted to provide overall antenna resonance at the operating frequency.

was no loss in the line.

What do you think happens in this example if your antenna is not matched properly? Suppose that your SWR meter shows a reading of 2:1. In this example the SWR at the antenna is really 5:1 and the total power loss is 5 dB. If your rig delivers 100 watts to this line, your antenna receives just over 30 watts. You've lost nearly a S unit (6 dB) on transmit and receive.

What if the line loss was 4 dB into a matched load? If the SWR meter showed a 2:1 reading, the SWR at the antenna would be over 11:1, causing an approximate 9-dB loss! Large line losses and high SWR do not mix well!

What if we used the same line on 40 meters? The loss into a perfectly matched load is about 1 dB. If your SWR meter indicates 2:1, then the SWR at the antenna is 2.5:1. The total loss is 1.3 dB. Barely discernible! This time 100 watts from the transmitter gives you about 75 watts at the antenna. Not ideal, but quite acceptable.

If you use open-wire line and a Transmatch, line losses are hardly a problem at all. If you experience a 0.1-dB loss with a perfect match while using roughly 100 feet of line at 10 meters, the losses total 0.5 dB with an SWR of 10:1. In this case, 89 of your 100 watts are delivered to the antenna feed point.

Shortened Antennas versus Full-Size Ones

Shortened antennas and trap antennas are attractive to those of us who live on small parcels of land. Some urban lots are as small as 60 by 60 feet, and this does present some perplexing problems for HF- and MF-band operators. We need, for example, to erect a 120-foot dipole, but the lot is only 60 feet on a side. The solution is to use a multiband trap vertical antenna, or a trap dipole (which is somewhat shorter than a full-size dipole at the lowest operating frequency (in MHz) to solve the problem. The dipole is usually erected as an inverted V.

What are the tradeoffs in performance for these compromise antennas? First, the best antenna performance will always come from a full size, single-band radiator. Secondly, trap multiband antennas have a much narrower SWR bandwidth than do full-size antennas. This condition worsens as the operating frequency in MHz is lowered. On 80 meters, for example, we might enjoy an SWR bandwidth of 100 kHz between the 2:1 SWR points, but with a trap antenna the 2:1 SWR bandwidth may drop to a mere 50 kHz.

The same performance tradeoff applies when we use loading coils to resonate an antenna (Fig. 1-3) that is less than 1/4 wavelength high or long. It makes no difference where the loading coil is placed in the antenna when we consider the decrease in efficiency. Loading coils introduce losses because they present an ac or RF resistance to the current flowing in the antenna. The smaller the loading-coil wire the greater the ac resistance of the coil.

Antenna traps act as loading coils on frequencies at which they are not resonant. Therefore, we may expect some losses in the system when traps are used. These losses are not particularly significant with respect to the advantages of having a shortened multiband antenna, but they are there, like it or not.

We can adopt a simple rule when thinking about shortened antennas: the shorter the radiator, the larger the loading coil, and the shorter the antenna the narrower the band-width. With the increase in loading-coil inductance, we will have higher losses in the system. It is possible to use, for example, a 2 foot vertical radiator on 160 meters, provided a large enough loading coil is used. But, such an antenna would radiate very little energy, and the 2:1 SWR bandwidth might be only 1 or 2 kHz! On the other hand, a 60-foot vertical for 160 meters (inclusive of a loading coil of much smaller proportions) could be used to work DX on that band, assuming a good ground system was employed. The bandwidth of the larger antenna should be on the order of 50 kHz, by comparison.

The Misunderstood Balun Transformer

There is probably no component about which more mumbo-jumbo is spoken or written. Few people seem able to even pronounce the word "balun" correctly. Pronunciation is bal, as in pal, and un, as in unhooked. Strangely, the word often comes out as "baylon, ballum or baloon." Balun is a derivitive of "balanced to unbalanced." This means that the device can be inserted between a balanced antenna feed point and an unbalanced feed line. This preserves the antenna balance to help ensure a nondistorted radiation pattern.

Baluns are broadband transformers. They are generally wound on a magnetic core, such as a ferrite rod or toroid. They can be constructed to provide an impedance transformation, such as 1:1 or 4:1. In the latter example we may use a 4:1 balun to convert a 200-ohm balanced-antenna condition to that of a 50-ohm unbalanced coaxial cable. If the balanced antenna feed point is 50 ohms, we can use a 1:1 balun for joining the antenna to a 50-ohm coaxial feeder.

You may hear all manner of balderdash about baluns, or you may find technical misstatements about them in ads and

magazine articles. One popular hyp is that baluns can cure
TVI. Or, you may read or hear that they can improve the
efficiency of your antenna. It is best to remember that
all transformers have some loss through them, however small
it may be. Losses are dependent upon how well the transformer
is designed and constructed. Also, if the balun or other
broadband transformer is not used correctly, the losses
will increase.

Correct Balun Use

A balun works properly when (1) it is terminated in the
correct resistive load. Specifically, a 1:1 balun should
have the same load impedance at both terminals. (2) The
balun must be used at some frequency within its design range.
(3) Baluns work best at impedances that are lower than 600
ohms. (4) A balun must be used within its power ratings,
and the greater the mismatch the less power it can handle
safely (without burning up or saturating).

Unfortunately, a balun may work as expected over a narrow
frequency range within a given amateur band, but may be
in a totally wrong environment in some other part of the
band. This is because the SWR may be 1:1 at the resonant
frequency of a matched antenna, but it may be 2.5:1 at some
other frequency within the band. When this event occurs
the balun no longer is able to perform its assigned function
in the classic manner.

It is true that a balun can help to ensure an undistorted
radiation pattern when we attach an unbalanced feeder to
a balanced antenna (such as a dipole), but for HF antennas
that are at low heights (where the directional pattern is
poorly defined), a balun will be of no benefit. Here we
are considering, for example, a 75-meter dipole that is
only 40 or 50 feet above ground. In the case of a Yagi beam
antenna for 20 meters (60 feet or more above ground), a
balun can, indeed, help to keep the radiation pattern uniform
(without skewing).

What Happens when the SWR is High?

The core material in a balun transformer will saturate if
the voltage and power through it is excessive for the cross-
sectional area of the core. The higher the voltage and power
(beyond the design limits of the transformer) the greater
the saturation. When saturation occurs the core will become
hot, change mu (permeability), and may even shatter. Before
this happens, as saturation commences, the core can generate
square RF waves, and this will cause harmonic radiation
(TVI and interference to other services). A saturated transformer core causes a reduction in the inductance of the
windings, and this renders the transformer useless. It can
actually cause the SWR to increase! High SWR is the usual

cause of the foregoing maladies. Therefore, we need to use our baluns in systems that are matched, or nearly so, if we want the balun to function as intended. I try to avoid using baluns in any system that has an SWR greater than 2:1. It can be seen from this that a balun can actually degrade an antenna system, rather than improve it.

Another problem that results from the use of a balun can be seen when the device is connected to an HF beam antenna that is designed for 20, 15 or 10 meters. Here, the leads from the balun to the driven element become a part of the driven element or matching device. This can alter the resonant frequency of the antenna (lower it), and it may disturb the matching device somewhat. Therefore, a balun should be installed directly at the antenna feed point, with the shortest lead lengths practicable. If you have experienced an increase in SWR when installing a balun, this is probably the reason why the SWR changed.

Baluns in Transmatches

You may be wondering why baluns have been indicted in the foregoing section. After all, the manufacturers use them in Transmatches to permit coupling to balanced feed lines! The same rules apply, no matter where the balun in installed in a feed line -- at the antenna or at the transmitter end of the circuit. Unfortunately, the instruction manuals for Transmatches with baluns often fail to contain information about the safe impedance range for the balun. Again, they should not be used when the impedance is much beyond 500 ohms, unless they are well insulated and have massive cores.

Transmatch instruction booklets often imply that the built-in balun can handle any matching situation associated with tuned ladder line or open-wire feeders. The fact of the matter is that under some conditions the feeder may present impedances of 1000 ohms or greater, depending upon the feed-line length and the type of antenna being used. A common example is seen in the case of a center-fed zepp antenna for multiband use. When the terminal impedance is high, there will be substantial RF voltage in the balun windings, and this can cause arcing, burning and saturation. Furthermore, the balun no longer functions as a broadband impedance transformer. It is because of these reasons that some manufacturers use teflon tape around the balun core, and use teflon-insulated wire for the transformer windings. This at least prevents the RF voltage from arcing between the windings and the core.

A much better arrangement for matching an unbalanced transmitter output to a balanced feeder is to use a balanced tuned circuit with coil taps for the open-wire line, such as is found in the E. F. Johnson Matchbox. The feed line is tapped on the coil at an equal number of turns from the

electrical center of the coil until a matched condition
is established. Since there is no balun in the line, arcing
and saturation can not take place, irrespective of the
impedance presented by the feed line, assuming that the
tuning capacitors and switches in the tuner are able to
accommodate the transmitter power output. See Fig. 1-4.

Capture Area or Aperture

Here is another misunderstood antenna concept. Many hams
believe that the larger the antenna the greater the "capture
area," and hence the better it will be for receiving signals.
The term "bigger" does not justify the claim of increased
aperture or capture area.

Capture area is a product of antenna gain. Therefore, if
we had a single dipole we would have a specific aperture
associated with it. If we erected a larger antenna with
the same gain, there would be no significant change in the
capture area. But, if we stacked a second dipole 1/2 wavelength above the first one, and fed them in phase, the gain
would increase 3 dB, and so would the aperture. We can see
from this that capture area is a function of antenna gain. It
follows that a 40-meter antenna, for example, that has gain
over a dipole, should have greater capture area.

Aperture is related also to frequency. For a specified power
gain the aperture is proportional to the square of the
wavelength. To illustrate this rule let's assume we have
a 600-MHz (0.5 m) antenna which has a power gain of 10 over
a dipole. A similar antenna with an identical power gain
at 148 MHz (2 m) would have a capture area 16 times greater
than the 0.5-meter antenna. If we were to obtain capture
areas of equal magnitude, we would have to increase the
gain of the 600-MHz array 16 times.

When antennas are stacked to increase the aperture they
should be arranged so that the apertures just touch. Too-
close spacing will cause the existing apertures to converge,
and this will reduce the effective area of the aperture.

The subject of aperture is far more complex than what we
have covered here. At best, the foregoing is a simplistic
treatment of aperture.

The Nitty Gritty of SWR

You may be more familiar with the term VSWR (voltage standing-wave ratio) than with SWR (standing-wave ratio), but
they stand for the same thing. Properly defined, SWR is
the ratio of the maximum voltage along a line to the minimum
voltage. The ratio of the maximum current to the minimum
current along the line is the same. Therefore, we may measure
the current or the voltage along a transmission line to

15

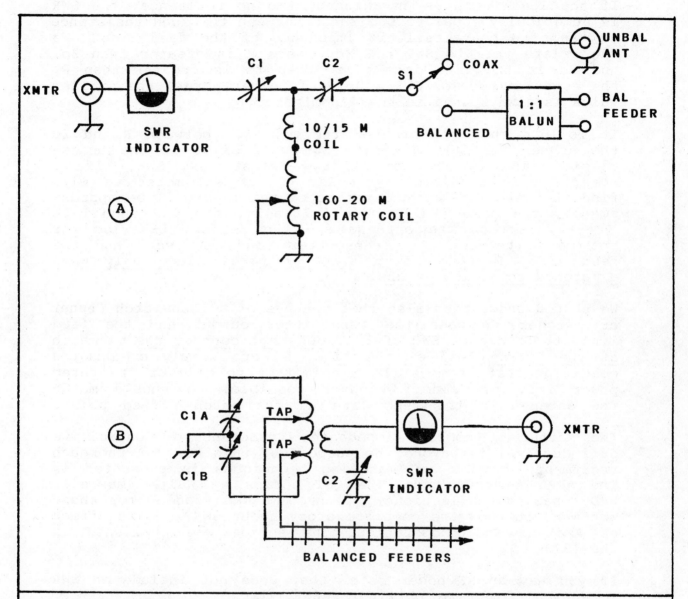

Fig. 1-4 -- The circuit at A represents a typical T-network
Transmatch with two variable capacitors and a rotary coil.
Many commercial tuners use this circuit, or a variation
thereof. The network is adjusted to obtain an SWR of 1:1.
A balun transformer at the tuner output is used for matching the transmitter to balanced feed line (see text). The
arrangement at B is better for working with balanced feed
line, because there is no need for the balun transformer
at A. The feeder is tapped along the tuned circuit to obtain
an SWR of 1:1.

determine the SWR.

If the load (such as an antenna) has no reactance, the SWR is numerically equal to the ratio between the load resistance (R) and the characteristic impedance of the feed line. We can relate this to SWR = R/Zo, where R is greater than Zo, and Zo is the line impedance. When Zo is greater than R, the equation becomes SWR = Zo/R. In other words, the smaller value is always used as the denominator.

It follows that the greater the mismatch between R and Zo the higher the SWR. Infinite SWR will exist if the transmission line is open or shorted (worst case SWR). If we observe this condition by means of an SWR meter we will find that the forward and reflected readings will be approximately the same (full scale). Ideally, the SWR should be 1 (1:1) between the antenna and the feeder. Likewise for the match between the transmitter (and receiver) and the feed line. Maximum power transfer takes place only when a matched condition exists.

We should understand also that the use of a Transmatch (tuner or matcher) between the transmitter output and the feed line, to obtain an SWR of 1:1, does not correct the mismatch at the antenna end of the line. We are simply creating a condition that permits the transmitter to deliver its rated power into the line. Whenever possible, we should match the antenna to the feed line at the antenna feed point.

The matter of complex impedances (reactance -- capacitive or inductive) is beyond the purpose of this text. A thorough treatment of this and matching techniques is presented in **The ARRL Antenna Book.** The QST series by Walter Maxwell, W2DU, called "Reflections," is recommended reading for those who want to know the complete story about SWR and its effect on Amateur Radio operation. The series was published in the 1970s.

If you have an RF power meter that does not include an SWR meter scale, you may obtain the SWR from

$$\text{SWR} = \frac{1 + \sqrt{\dfrac{RP}{FP}}}{1 - \sqrt{\dfrac{RP}{FP}}} \qquad \text{Eq. 1-1}$$

where RP is the reflected power and FP is the forward power, as indicated on the power meter in watts or milliwatts. You should find the nomograph of Fig. 1-5 useful for computing the SWR when you are working with an RF power meter.

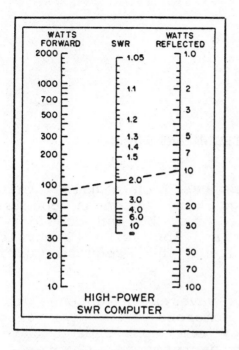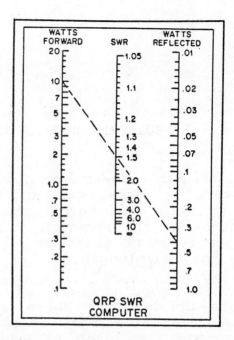

Fig. 1-5 -- Handy nomographs for determining the
SWR when the forward and reflected powers are known.
(Courtesy ARRL, Inc.)

The Isotropic Radiator

You will hear about an isotropic antenna during your Amateur
Radio experience, and you may be confused about the meaning
of this term. In simple language, the isotropic antenna
is an ideal imaginary radiator in free space. Think of it
as a perfect sphere that radiates equally in all directions.
No such practical antenna exists, and if it did we would
find it impossible to feed energy to it without disturbing
the perfect radiation pattern.

The concept of an isotropic radiator is useful for establishing a reference against which to compare the performance
of practical antennas. If such an antenna existed it would
have no directional properties whatsoever.

Meaningful antenna-gain measurements are generally compared
to the characteristics of a reference dipole. In other words,
a three-element Yagi beam antenna might have 7.5 dB of gain
(in its favored direction) over a dipole, but the figure
would be substantially higher if referenced to an isotropic
radiator. Some manufacturers like to use the isotropic source
as a reference in their ads. This inflates the actual gain
in dB, and such claims can mislead the unwary buyer. The
reference is rendered as dBi, for decibels over isotropic. To
summarize this discussion, we can think of the isotropic
radiator as a purely hypothetical device that is useful
for scientific comparisons when evaluating antennas that
have various directional properties.

Chapter 2

BUILDING and USING DIPOLE ANTENNAS

Most of our amateur antennas are based on a half-wavelength of wire or metal tubing. Large wire antennas consist of two or more half wavelengths of wire. We can consider such antennas as the rhombic, V beam and long wire (1 wavelength or greater) as examples of systems that require multiples of a half wavelength.

The most popular form of half-wavelength antenna is the dipole or doublet. There are numerous forms for dipoles. We will examine them, how they are constructed and what to expect from each by way of performance. Chapter 1 has information about antenna height versus performance. Please read that section before you erect your dipole.

Free Space versus Real-World Dimensions

If we could place a half- or full-wavelength antenna in free space (the ionosphere, for example), it would have a physical dimension that is somewhat longer than that of the same antenna erected near earth. This is because the earth and conductive objects near the antenna (such as trees, buildings and power lines) present stray capacitance that tunes the antenna lower in frequency than the free-space equivalent antenna. You may think of an antenna as an inductance (coil) and capacitance in parallel (a tuned circuit). The nearer your antenna is to conductive objects the shorter it must be for resonance at the desired operating frequency.

The correct length for a half-wavelength antenna in free space may be obtained from 492/f(MHz) = feet. From this we learn that a half wavelength at 3.8 MHz is 129 feet and 5-1/2 inches. The same antenna erected on your property would be approximately 123 feet, 2 inches, based on the approximate calculation of 468/f(MHz) = feet. The 468 number has been used for many years to determine the initial length of a half wavelength. Depending upon your installation, you may have to shorten or lengthen the antenna slightly to make it resonant and to secure the lowest SWR at the frequency of operation. We can see from this that there is no magic equation for cutting our antenna wires to a precise length.

Once your antenna is in place on its supporting mast, tree or tower, you may trim it for resonance. This is done with the aid of an SWR indicator or an antenna noise bridge. Small amounts of wire are added or removed (equally from each side of the dipole) until you obtain the lowest SWR in the center of the proposed antenna operating range, say, from 3.7 to 3.9 MHz for an 80-meter dipole. Fig 2-1 shows various forms for dipole antennas.

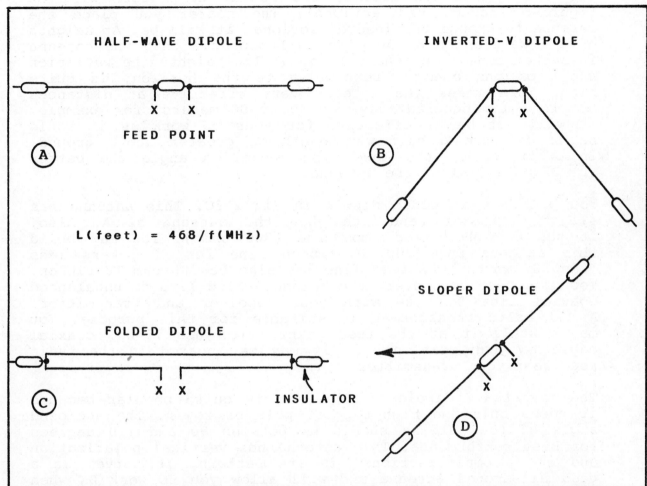

Fig 2-1 -- Popular formats for half-wave dipole antennas. The example at A provides horizontally polarized radiation, and the directivity is bidirectional off the broad side of the antenna when it is high above ground. Antenna B is an inverted-V or drooping dipole. The pattern is omnidirectional with vertical polarization. A folded dipole is shown at C. It has the same characteristics as dipole A, but has a higher feed-point impedance (300 ohms). The other antennas have a feed impedance between 50 and 75 ohms typically, depending upon the height above ground. Antenna D is similar in performance to the inverted V when supported by a non-conductive mast. A metal mast or tower causes maximum directivity off the slope of the antenna (arrow).

Dipole Performance

Each of the dipoles in Fig 2-1 is one half wavelength overall.
Each leg of the dipole is 1/4 wavelength long. The feed
line is attached at the exact electrical center, as shown
by the designators "X." If the antennas at A and C of Fig
2-1 are high above ground, the radiation pattern will be
figure-8 shaped, bidirectional. The nearer you place the
antenna to ground the less directional it will be. At heights
of 1/4 wavelength or less you will observe that the antenna
is rather omnidirectional. Also, at low heights the radiation
angle of the energy, respective to the horizon, is quite
high. This makes the antenna very effective at distances
out to a few hundred miles on 40 or 80 meters, for example.
It will not be as effective for your DX work as it would
be if it were a half wavelength or greater above ground.
In other words, the lower the radiation angle the better
the DX potential of the antenna.

You will see a folded dipole in Fig 2-1C. This antenna has
slightly greater bandwidth than the antenna at A. Also,
it has a higher feed impedance (300 ohms). You may build
this antenna from 300 TV ribbon line for RF powers less
than 300 watts. The feed line may also be 300-ohm TV ribbon.
You will need to convert the balanced feed line to unbalanced
coaxial line for use with your receiver and transmitter.
A 4:1 balun transformer is suitable for this purpose. You
may install it at the feed point (then use 75-ohm coaxial
cable to your station), or it may be located in the radio
room near your transmitter.

The inverted-V dipole of Fig 2-1B is quite popular because
it needs only one high support pole or tower. The enclosed
angle of the antenna should be between 90 and 110 degrees
for best performance. This antenna has vertical polarization
and it is omnidirectional in its pattern. It serves as a
good all-around antenna and will allow you to work DX when
band conditions are favorable.

Fig 2-1D shows a sloper or sloping dipole. This is sometimes
called a "full sloper." It offers the same advantages as
the inverted V (vertical polarization and omnidirectional)
when it is supported by a nonconductive mast. When it is
supported from a metal mast or tower it has maximum directivity off the slope of the dipole. This is indicated by
the arrow. Despite this directivity, the antenna is suitable
for operation in all compass directions. The greatest null
in the radiation pattern will be off the side of the mast
or tower opposite the dipole.

Dipoles are used for limited-distance communications when

they are only 10 or 15 feet above ground. This is not a
wise thing for you to do if you want to work stations beyond
a radius of 200 to 300 miles on a regular basis, especially
on 160, 80 and 40 meters.

Dipole Construction

You should have a well-constructed center insulator for
your dipole. It should be constructed to relieve the strain
on the feed line where it is attached to the feed terminals.
There should be no rough edges on the center block where
the legs of the dipole come in contact with it. Rough edges
cause stress on the wire during periods of wind. This can
lead to broken wires. The same rule is valid for the dipole
end insulators. The openings to the holes need to be smooth
and free of burrs. The longer the end insulator the more
effective it will be in isolating the dipole from the guy
wires (desirable). This is especially important when there
is dirt, rain, snow or ice on the insulators. High-quality
insulating material should be used for the end insulators.
Materials such as steatite, ceramic, Teflon, glass, Delrin
and polyethylene are especially good for RF insulation.
The center insulator of the dipole need not have high quality
insulating properties, although this is a desirable objective, nonetheless. The dielectric quality of the center block
may be lower because the dipole center is low impedance
(single-band dipole). Therefore, high values of RF voltage
are not present in this part of the antenna. Maximum RF
current will be present instead. The high RF voltages appear
at the ends of the dipole, hence the need for good insulation at those points. Fig 2-3 shows how to construct a good
center block from plastic material.

The principal drawing in Fig 2-3 is for a center insulator
to be used with a straight, horizontal dipole. The smaller
inset drawing shows how you can change the center-block
design to accommodate an inverted-V dipole. The top hole
is for the supporting guy line that holds the antenna aloft
on a mast or tower.

Stranded no. 12 or 14 copper wire is best for dipole-antenna
construction. It can withstand considerable flexing before
it breaks. This is not true of single-strand copper wire
of equivalent gauge. The exception is when you use copperweld wire (steel core). Unfortunately, copperweld is like
spring stock, and is therefore difficult to work with. Also,
sharp bends in copperweld cause cracks in the copper coating.
This leads to ultimate breakage when the steel core rusts.

The antenna wires in Fig 2-3 need to be soldered securely
to the solder lugs. This prevents intermittent operation
and eliminates the chance for stray rectification at the
connections (TVI and RFI).

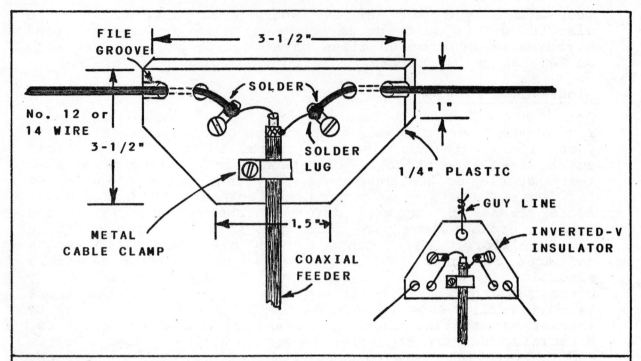

Fig 2-2 -- Pictorial representation of a center insulator for a dipole antenna. This center block can be made from sheet PVC, high-impact polystyrene sheet or similar high-strength plastic, such as Lexan. A metal cable clamp is used to affix the coaxial feeder to the block. The open end of the cable should be sealed against dirt and moisture by means of epoxy cement or Coax Seal. A half-round channel needs to be filed at each side of the block where the antenna wire joins the block. A mouse-tail file is suitable for this. The channel prevents the sharp edge of the block from cutting into the wire. The small drawing shows how to fashion an inverted-V center block from plastic.

Multiband Dipoles

The dipole antennas in Fig 2-1 are for single-band use. You have the choice of erecting several single-band dipoles, or you may use one dipole and tune it for various HF bands. This technique is less expensive than when two or more dipole antennas are constructed. The major saving is in the feed line. A multiband dipole needs but one feeder. The multiband dipole does, however, require the use of a Transmatch to tune the system at the operating frequency. Depending upon the type of Transmatch you buy or build, you may also need to use a balun transformer. This converts the balanced feeder to unbalanced coaxial line. The balun is used between the balanced feed line and the Transmatch. Some Transmatches have a balanced tuning system, such as the E.F. Johnson Matchbox, which eliminates the need for a balun transformer. Fig 2-3 shows the two types of Transmatch, along with the circuit for a multiband dipole with tuned feeders. This antenna is sometimes referred to as a center-fed Zepp.

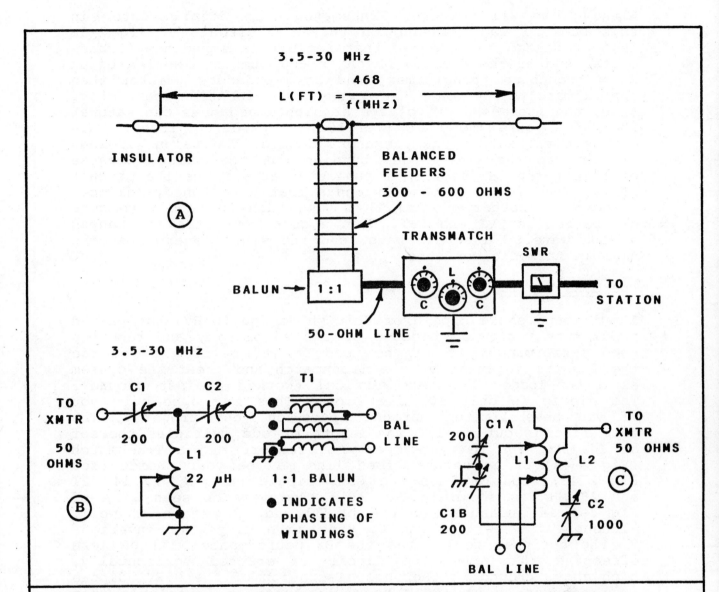

$$L(FT) = \frac{468}{f(MHz)}$$

Fig 2-3 -- Example of a multiband dipole. It is cut for the lowest desired operating frequency (A). Feed line may be kW 300-ohm TV ribbon, 450-ohm ladder line or homemade open-wire line. The feed line can be any length, but best results are claimed by those who use a 1/4-wave feed line for the lowest operating frequency. A 1:1 balun transformer converts balanced feed to unbalanced line. The balun is generally located within the Transmatch and may be subject to heating and arcing when the feed line presents a high impedance to the balun. The Transmatch is adjusted to obtain a 1:1 SWR. Diagram B shows a typical T-network Transmatch. A balun is used for balanced feeders. It is bypassed for use with coaxial feeders or end-fed wire antennas. The Transmatch at C is best for multiband antennas with tuned feeders. The feeders are tapped on L1 to obtain an SWR of 1:1. C1 and C2 must also be adjusted to obtain a low SWR. A balun is not needed with this Transmatch. L1 must also have taps for decreasing the inductance for operation on the higher bands. A switch is normally used for this purpose. A balun transformer is not required for the circuit at C.

Although we will discuss Transmatches in detail, later in
this book, I want to mention that the circuit of Fig 2-3B
is best suited to antennas that reflect an impedance (transmitter end of the feeder) which is 600 ohms or less. A balun
is a broadband transformer, and at impedances greater than
approximately 600 ohms it ceases to function correctly.
Also, the increased RF voltage at high impedances can saturate
the balun core material (ferrite or powdered iron). This
causes core heating and permanent damage. The balun windings
may arc to the core, thus burning out the balun. Because
of this potential problem I suggest that you use the circuit
of Fig 2-3C for multiband antennas that have tuned, balanced
feeders.   Another problem that goes hand in hand with core
saturation is TVI and RFI. When a core saturates it changes
a sine wave to a square wave, and the signal energy becomes
rich in harmonic currents.

The G5RV Dipole

A multiband short dipole of British design (G5RV) has gained
popularity worldwide because of its simplicity and reported
good performance. It may be used on 160 meters if you tie
the feeders together at the Transmatch and treat the system
as a top-loaded 1/4-wave vertical (worked against ground).
The dipole is only 102 feet long. This may appeal to you
if you have limited antenna space on your property. The
feed line is cut to 1/4 wavelength or odd multiples thereof
at 14 MHz. You may use open-wire feeders and a Transmatch
(Fig 2-4A) or an untuned feed line may be your choice (see
Fig 2-4B). You may use the G5RV dipole on 3.5, 7, 14, 21
and 28 MHz as shown in Fig 2-4B. The version seen at A of
Fig 2-4 is suitable for use on 30 and 12 meters as well.
Although this antenna can be used as a sloper or inverted-
V, the designer feels that the DX performance will be less
effective than when the dipole is erected horizontally,
high above ground. A thorough discussion of the G5RV dipole
is presented in the RSGB book, HF Antennas (available from
The ARRL), chapter 11.

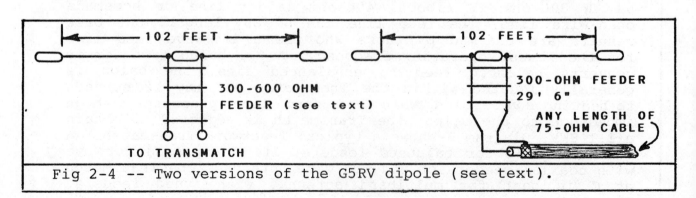

Fig 2-4 -- Two versions of the G5RV dipole (see text).

A Fanned Multiband Dipole

You may attach several dipoles to a single coaxial feed line to obtain multiband operation. Each dipole is cut for a specific amateur band. The individual dipoles are fanned apart to minimize interaction. This practice makes it easier to adjust the individual dipole lengths. When they are close to one another, each dipole tends to detune the nearest ones, and the lengths will be somewhat different than when a single dipole is dimensioned from 468/f(MHz). Fig 2-5 shows how you may construct a fanned, multiband dipole.

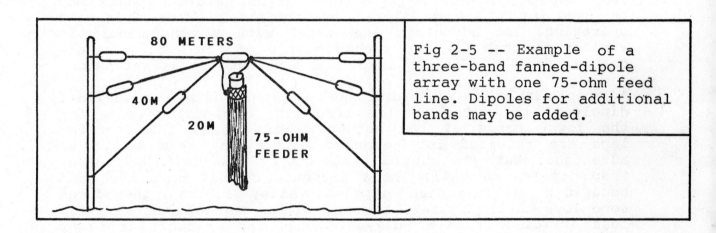

Fig 2-5 -- Example of a three-band fanned-dipole array with one 75-ohm feed line. Dipoles for additional bands may be added.

You will discover that these dipoles are somewhat more tricky to adjust for resonance than is a single-band dipole. This is because of the interaction I mentioned previously. The best way to deal with the problem is to adjust the longest dipole first. It will be affected the least by the presence of the remaining shorter dipoles. Next, adjust the dipole for the higher adjacent band (80 to 40 meters, for example), and so on until each dipole is resonant in your favored part of each band.

A word of caution is in order at this time: Any multiband antenna that does not require a Transmatch will radiate harmonic energy effectively. If your transmitter is not reasonably free of harmonic energy (poor design or defective), you should use a harmonic filter between the transmitter and the feed line. When a Transmatch is used with any antenna system, some harmonic attenuation should occur by virtue of the tuned circuit in the Transmatch. The magnitude of the harmonic attenuation is dependent upon the tuner design and the specific settings of the controls for a given condition of matching. The Transmatch of Fig 2-3C offers the greatest amount of harmonic attenuation, compared to the network of Fig 2-3B.

Understanding Trap Dipoles

You can build a multiband dipole that contains traps. This kind of radiator is called a "trap dipole." It is not a "trapped dipole" (misnomer), as you may hear it referred to by other hams. Specifically, the antenna is not <u>trapped</u>. Rather, it contains traps. You may think of an antenna trap as a barrier against the flow of RF current. Traps consist of a coil and a capacitor in parallel (Fig 2-6A). This forms a tuned circuit that presents a high impedance to the RF current for the frequency of the trap. A trap "divorces" the portion of the antenna beyond the trap. The feed line effectively "sees" only that part of the antenna out to the trap of interest. Traps for various amateur bands may be used in a single dipole antenna to provide multiband operation. The advantage associated with a trap dipole is similar to that for the fanned dipole of Fig 2-5: A single coaxial feed line is required, and a Transmatch is not needed.

A trap dipole is shorter than a single-band conventional dipole. This is because the traps act as loading coils for the lower bands of the antenna. This means that shorter legs are required for resonance. This may seem to be an advantage, but you should be aware of the deficiencies of trap dipoles as well. These antennas do not have the same bandwidth as full-size dipoles. Also, a trap introduces some loss in the system, thereby making the antenna slightly less efficient than a full-size one. The tradeoff is minor compared to the advantages of the trap antenna. The loss is so slight (if well-designed, high-Q traps are used) that other operators may be unable to see the signal difference on their S meters. The term "Q" stands for quality factor. If a tuned circuit or other component has a high Q, this means that the losses are minimal, and that the device has minimum unwanted ac or RF resistance in the signal path. High-quality insulation goes hand in hand with high Q. The narrower bandwidth associated with trap antennas results in a smaller 2:1 SWR bandwidth for each band below the highest one for which the antenna is designed. For example, a full-size 80-meter dipole might have a bandwidth of 150 kHz between the 2:1 SWR points. Conversely, a trap dipole for 80 and 40 meters may have an 80-meter bandwidth of 75 kHz.

You will see other types of trap antennas in use by fellow amateurs. Some use trap verticals (one half of a dipole), while others may have a trap Yagi beam antenna on a tower. In any event, the principle of trap operation and antenna bandwidth remains the same as for a trap dipole. In other words, there are advantages along with a tradeoff in antenna performance. You will observe also that trap antennas are substantially heavier than non-trap dipoles. This means that additional care is needed in supporting them.

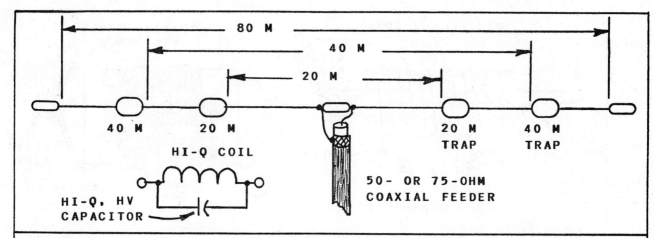

Fig 2-6 -- A trap dipole that contains tuned circuits of the type seen in the inset drawing. These dipoles may have as many traps as needed to provide multiband coverage. The example above is for operation on 80, 40 and 20 meters. Each trap is tuned to the center of the desired operating range of each band. Trap tuning is done by means of a dip meter before the traps are installed in the antenna. The traps must be protected from dirt and moisture. Plastic cylinders, such as polyethylene food containers, can be used as protective covers.

Trap Construction

Your homemade traps may be fashioned from sections of Mini-ductor coil stock (B&W Co.) and ceramic, HV transmitting capacitors. Choose coil stock with heavy-gauge wire, such as no. 12 or 14. This will help to ensure high Q and minimum losses. Although a high-quality coil form can be used for winding a trap coil, try to use air-wound inductors. This will keep the trap weight low while minimizing dielectric losses. Moisture and dirt have a greater degrading effect on traps that are wound on coil forms, as opposed to air-wound ones.

Another type of trap is the kind that is made from small coaxial cable. The coaxial line takes the form of a close-wound coil on a low-loss form. The coaxial line provides inductance and capacitance. This eliminates the need for a separate fixed-value capacitor. Fig 2-7 shows both types of trap. A comprehensive article about coaxial-trap design was published by N4UU in Dec. 1984 QST, page 37.

Coaxial-line traps are adjusted to resonance by pruning the coil turns while checking the frequency with a dip meter. This can be a tedious job with this variety of trap. Precise winding data for each HF band is given by N4UU in his QST article. He shows how many turns of cable are needed for various coil-form diameters. PVC pipe may be used for the coil forms, but thin-wall, high-impact polystyrene tubing is lighter and has better insulating properties (check with your local plastics dealer about this tubing).

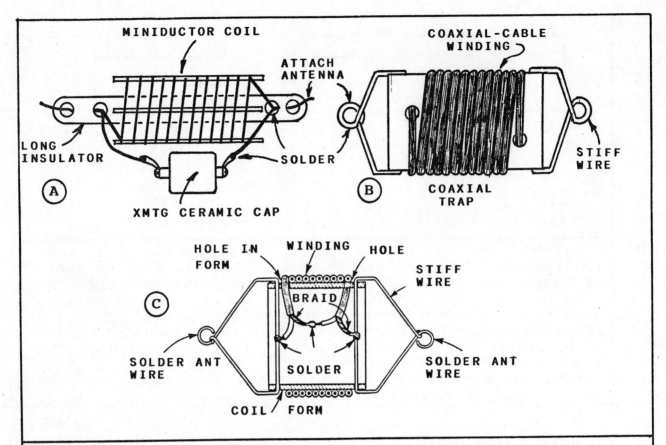

Fig 2-7 -- Pictorial drawings of a coil/capacitor trap (A) and a coaxial-cable trap (B). A breakaway view of the coaxial trap is shown at C. Note that the shield braid from one end of the winding is connected to the center conductor of the opposite end of the winding.

You may be wondering how to select the proper values of capacitance and inductance for the trap shown in Fig 2-7A. A suitable rule of thumb calls for a reactance of approximately 400 for the L and C components. The component values may now be determined from

$$C(\mu F) = \frac{1}{6.28 \times f(MHz) \times X_c} \quad \text{Eq. 2-1}$$

$$\text{Example: } C(\mu F) = \frac{1}{6.28 \times 7.1 \times 400} = 0.000056 \text{ or } 56 \text{ pF}$$

To find the required inductance use

$$L(\mu H) = \frac{X_L}{6.28 \times f(MHz)} = 8.97 \text{ } \mu H \text{ for 7.1 MHz} \quad \text{Eq. 2-2}$$

As we learned earlier, final trap adjustment can be done by adjusting the coil turns while checking the trap with a dip meter.

Dipole Look-Alikes

You will see long antennas that resemble the half-wave
dipole in structure. By proper definition, they are not
classic dipoles. The IEEE Standard Dictionary defines a
dipole as, *a metal radiating structure that supports a
line current distribution similar to that of a thin straight
wire, a half wavelength long, so energized that the current
has two nodes, one at each of the far ends*. The language is
specific about a half wavelength being the electrical size
of the radiator.

We can build two-wire antennas that resemble a dipole,
and they will provide gain over a dipole. You may wish
to consider these antennas if you have room for them on
your property. Perhaps the most popular of these larger
antennas is the collinear array that consists of two half
waves in phase. This antenna is shown in Fig 2-8. It has
two half-wavelength wires, and the feed line is attached
at the inner junction of the wires, as is the case with
a dipole. It is customary to feed this antenna with balanced,
open-wire feeders. You may also use 300-ohm TV line or
450-ohm ribbon or ladder line, but the open-wire line is
best in terms of minimum line loss. The feed impedance
is on the order of 4000 to 6000 ohms (theoretical), which
means that the feeders will cause a mismatch. This is not
a vital consideration, owing to the very low loss in open-
wire feed line. This principle is associated also with
multiband dipoles that are fed with balanced feeders. In
the case of a dipole, the feed impedance is low (50 to
75 ohms, approximately) in the lowest operating band, but
it becomes very high on the harmonic frequencies. By way
of an example, a dipole that is designed for 80 through
10 meter use will exhibit a 50- to 75-ohm feed impedance
at 3.5 MHz, but on 40 meters it becomes two half waves
in phase (collinear array), and the feed impedance rises
to several thousand ohms.

The theoretical gain of this two-element collinear antenna
is 1.9 dB over a dipole. This means that if your transmitter
puts out 64 watts, the effective radiated power from the
antenna rises to 100 watts in the direction of maximum
radiation. A two-element collinear array has a radiation
pattern similar to that of a dipole -- broadside to the
antenna, with a figure-8 pattern. Once again we must be
aware that the foregoing conditions can only prevail if
the antenna is erected approximately 1/2 wavelength above
ground, or greater.

You may obtain additional gain from a collinear array by
adding half-wave sections and quarter-wave stubs, as shown
in Fig 2-9. The stubs are necessary in order to obtain
the required phase reversal between the half-wave sections.

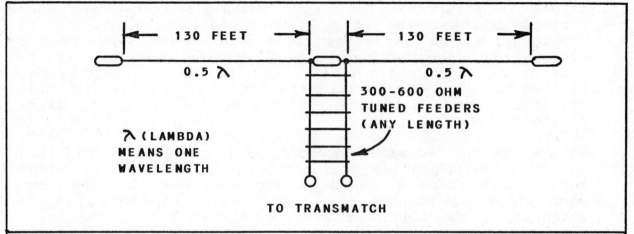

Fig 2-8 -- Two half-wave wires may be combined as shown to form a two-element collinear array. The theoretical gain of this antenna over a dipole is 1.9 dB. It has a bidirectional radiation pattern off the broad side of the antenna. Tuned, balanced feeders are required. The diminsions given are for operation at 3.6 MHz.

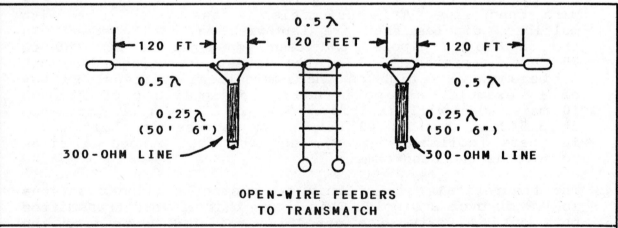

Fig 2-9 -- Three or more half-wave wires can be combined to form a three- or four-element collinear array. This example shows how three half waves are arranged. Quarter-wave phase-reversal stubs (two) are required for a collinear array of this kind. The velocity factor of the line used for the stubs must be included in the calculations for the stub length. This example is for operation at 3.9 MHz. The theoretical gain of this antenna is 3.2 dB.

The three-element collinear array of Fig 2-9 has an approximate gain of 3.2 dB. This means that the effective radiated power (ERP) of a 100-watt-output transmitter is 210 watts in the directions of maximum antenna radiation. A four-element collinear array provides roughly 4.3 dB of gain over a dipole. Four collinear elements appear to be the point of diminishing return, respective to increased array gain.

You may wish to consider the system in Fig 2-9 for use on 40 meters. The diminsions would be shortened accordingly. Such an antenna would be useful on 160 and 80 meters also, since the phasing stubs would act somewhat as loading coils to make the antenna sufficiently long for the two lower bands. The system would then function as a nonresonant antenna with tuned feeders.

It is important to minimize the tuned-feeder length as much as is practicable. This will reduce system losses. You may prefer also to use open-wire line for the phasing stubs. The velocity factor for TV ribbon line is 0.80, whereas it is 0.95 for open-wire line. Therefore, the stubs will be slightly longer if you use open-wire line for your lines. You may also use 450-ohm molded ladder line. The approximate velocity factor for this material is 0.85.

Long, high-quality insulators should be used in collinear arrays. This is because high values of RF voltage are found at each of the insulators. Good isolation between the ends of the collinear elements is necessary in order to ensure proper phasing, especially during wet weather. The reversal of the phase should occur only in the stubs.

Chapter Summary

I have highlighted only the practical aspects of dipole and similar antennas. You will have no trouble duplicating and tuning the antennas discussed in this chapter. They represent the most common of the dipole antennas used by amateurs. With the exception of the collinear arrays of Figs 2-8 and 2-9, all dipoles can be used as shown in Fig 2-1. The choice of horizontal, sloper or inverted-V antenna is entirely yours to make. It will depend upon what type of performance you desire. Part of your judgement will be founded also on available property and supporting structures.

The fundamental rules to keep in mind are (1) erect your antenna as high as possible. (2) Keep it as far from conductive objects as you can. (3) Use low-loss feed line, and keep the feeders short to reduce losses. (5) Solder all antenna joints. (6) Use high-quality insulators.

Chapter 3

SINGLE-WIRE ANTENNAS

If you think about antennas as I once did, you are tempted
frequently to string up a cheap and simple end-fed wire
antenna. Chances are that it will not be very high above
ground, since that often simplifies the installation. I
don't wish to pour cold water on your enthusiasm for end-
fed antennas, but they may cause some problems (along with
poor performance) that you do not need nor want!

I would be unfair and incorrect if I were to say that we
should always avoid the use of end-fed antennas. There
are times when, owing to physical restrictions, we have
no choice other than this style of radiator.

It would be grossly unfair to indict the end-fed antenna
as a black sheep among other wire antennas. You and I have
used these antennas at least a few times when the quality
of performance was outstanding, all things considered.
Good results are contingent upon antenna height, wire length,
proper impedance matching and the quality of the ground
below the antenna. Conductive clutter near the antenna
has the same effect on end-fed wires as it does on dipoles
and other radiators. So, performance is based also on the
near environment of the antenna. Because of these consider-
ations, a particular end-fed antenna may produce superb
results at one location: At some other location it may be
on par with an air-cooled dummy load. It is necessary to
experiment with various antennas in order for us to learn
which type does the best job at a given location.

Some Problems With End Feed

You should be aware that when you employ an end-fed antenna,
part of the radiator is usually inside your ham shack.
This means that there can be an intense RF field around
your station equipment. This frequently causes the mic,
key and equipment cabinets to be "hot" with RF voltage.
Not only will this unwanted energy "bite" you when you
touch the station apparatus, it can (and often does) get
into audio, VFO and keyer circuits to cause serious mal-
functioning. The foregoing is the curse associated with
end-fed wires!

The better your station ground system the less trouble
you may have with stray RF energy. This is true, at least,
for a single operating frequency. But, it is not uncommon
to have an "RF-cold" shack on one band, only to find that
RF energy is rampant on other amateur bands.

The difference between antenna conditions from one band
to another can be blamed on how effective the grounding
system in the shack is at a specific frequency. Generally,
you will have your greatest problem with stray RF energy
at the higher frequencies of the HF spectrum. A long ground
lead allows standing waves to appear on it, especially
at 7 MHz and higher. The magnitude of the RF energy in
your shack will depend also on whether the equipment end
of your wire antenna presents a current or voltage node
in the shack. You will have the least trouble when the
end of the wire in your house is low impedance (current
fed, or a voltage node). This will be the situation if
your wire is 1/4-wavelength long, or an odd multiple thereof.
The worst problems occur when the wire is 1/2-wavelength
long, or a multiple thereof. This places a current node
at the rig (high RF voltage).

Random-Length Wires

You will hear some amateurs say that they are using a
"random antenna." This should indicate that they don't
know exactly how long the wire is. Chances are that they
strung an unknown length of wire between two available
supports, and brought one end of the wire into the radio
room. If the wire is less than 1/4 wavelength overall,
it will present capacitive reactance (X_c) to the transmitter.
This capacitive reactance will need to be cancelled by
inserting an inductance of equal reactance in series with
the feed point (loading coil) to make the antenna look
like a resonant quarter wavelength, respective to the transmitter and receiver. This is illustrated in Fig 3-1A.

Suppose, on the other hand, that the wire is slightly longer
than 1/4 wavelength. This causes the feed point to exhibit
inductive reactance (X_L). Compensation requires a capacitive reactance of an equal value (in series with the antenna)
to cancel the inductive reactance. This does not mean that
the reactance compensation (Fig 3-1B) will cause the feed
impedance to become 50 ohms. Rather, the feed point (end
of the wire) will be resistive at the compensated frequency.
A resistive condition occurs at antenna resonance, but
the feed impedance may be any value, such as 15 or 100
ohms. This means that you need to use a matching network
(Transmatch) to obtain a 1:1 SWR for your 50-ohm equipment.

The correct way to ensure a 50-ohm resistive feed impedance
for your end-fed wire antenna (random length) is illustrated
in Fig 3-1C. Most Transmatches will, however, provide a
50-ohm match without using the C or L reactance compensator
shown in Fig 3-1A and B. The Transmatch will operate better
(tune easier within its range) if you make the antenna
resistive (Fig 3-1A and B) before you use the Transmatch
to provide an SWR or 1:1. This discussion applies   to

single-band operation. If you use the same antenna for
multiband operation, the C and L reactance components will
serve no useful purpose. You will fare better by simply
connecting the end of the random-length wire to your Transmatch, then tuning the system for a 1:1 SWR.

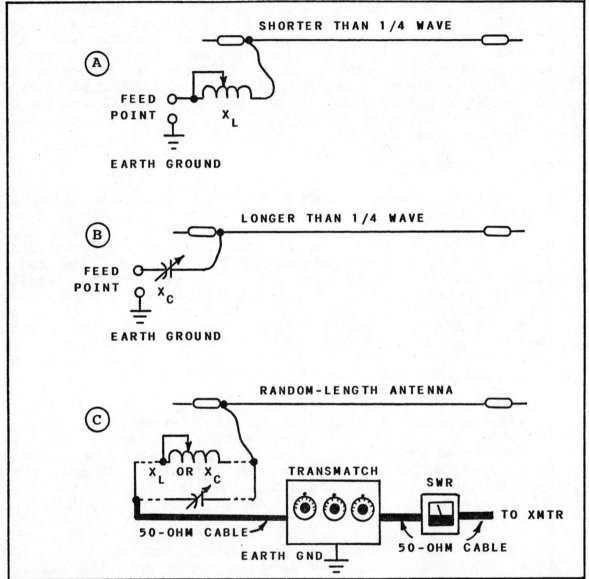

Fig 3-1 -- Examples of reactance compensation for wires
less than (A) or more than (B) 1/4 wavelength long (see
text). Illustration C shows how to use L or C reactance
compensation along with a Transmatch and an SWR meter. Each
example shows a random-length, end-fed wire. X_L is a roller
inductor or a tapped coil to permit a wide range of adjustment for unknown values of X_C at the feed point. Variable
capacitor X_C requires sufficient plate spacing to prevent
arcing.

Random-Wire Matching Procedure

Although we have discussed and illustrated (Fig 3-1) how
to tune out the reactance of a random-length antenna, let's
recognize, first, the need for a good ground system against
which to work the electrical 1/4-wavelength wire. A ground
screen or counterpoise, plus the station earth ground,
is necessary in order to obtain maximum performance from
this antenna. The ground screen or counterpoise should
be attached to the transmitter and/or Transmatch metal
chassis.

Assuming we have satisfied the ground-system requirement
for our antenna, we may proceed with resonance and matching
adjustments. The reactance compensators (L or C) of Fig
3-1 may be adjusted by placing an SWR indicator between
the transmitter output and the rotary inductor or variable
capacitor. These components are adjusted (transmitter on)
for the lowest SWR attainable. This will indicate antenna
resonance, and the feed point will be purely resistive
at this frequency. It is possible that the SWR will be
less than, say, 2:1 without adding a Transmatch. If this
is the case, you may operate without the tuner. If, however,
the impedance is such that the SWR is greater than 2:1,
add the Transmatch, as shown at C of Fig 3-1. It can be
adjusted to provide an SWR of 1:1.

End-Fed Zepp Antenna

A true Zepp (Zepplin) antenna is fed at one end. Although
we hams refer to a center-fed dipole (tuned feeders) as
a "center-fed Zepp," this is a misnomer. The principle
of operation is, however, similar for both antennas.

An end-fed Zepp consists of a half-wavelength radiator
that is feed at one end by means of tuned feeders. This
causes an imbalance in the feed system, even though balanced
feed line is used. One conductor of the feeder is open,
so to speak. This causes standing waves on the feed line.
This situation causes the feeders to radiate, along with
the half-wavelength wire. This will not lead you to have
serious problems, but the feeder radiation will cause the
antenna radiation pattern to be distorted. There will be
vertical and horizontal radiation from the system, provided
the feed line is perpendicular to earth.

The principal advantage of using tuned feed line with your
end-fed antenna is that the voltage end of the antenna
is no longer in your ham shack. The impedance at the shack
end of the tuned feeders should be fairly low (current)
if you use a quarter-wave (or odd multiple thereof) feed
line from the end of the wire to the Transmatch. This helps

prevent RF energy from appearing   on your mic, key and
equipment cabinets. An end-fed Zepp is shown in Fig 3-2. An
end-fed Zepp can be used as a multiband antenna by employ-
ing a Transmatch. Be prepared, if you do this, to discover
unwanted RF energy in your shack on certain ham bands,
since the feeder will present a high impedance (RF voltage)
at the Transmatch, owing to its being a half-wavelength
long (or multiple of a half wavelength) at the higher frequencies.

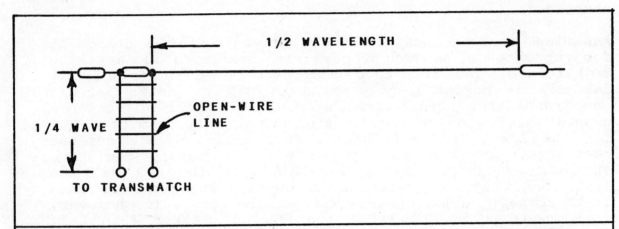

Fig 3-2 -- An end-fed Zepp antenna uses tuned feed line.
Minimum unwanted RF voltage will appear in the station
if 1/4-wavelength feeders are used. Some feed-line radiation
will occur because one side of the feed line is unterminated.

Long-Wire Antennas

An antenna is only a <u>long wire</u> if it is one wavelength
or greater overall. A physically long piece of wire does
not constitute a "long wire," contrary to what you may
have heard people call them. Length in this case relates
to <u>electrical</u> dimensions of 1 wavelength or more.

A long wire may be fed at one end, as recommended for the
Zepp antenna in Fig 3-2. Alternatively, you may locate
the feed point 1/4 wavelength from one end of the wire
to allow the use of low-impedance coaxial transmission
line. Illustrations of both methods are shown in Fig 3-3.
The length of a long wire is dependent upon the number
of wavelengths it contains. The following formula may be
used to find the antenna length:

$$L \text{ (feet)} = \frac{984 (N - 0.025)}{f \text{ (MHz)}} \qquad \text{Eq. 3-1}$$

where N is the number of wavelengths in the antenna.

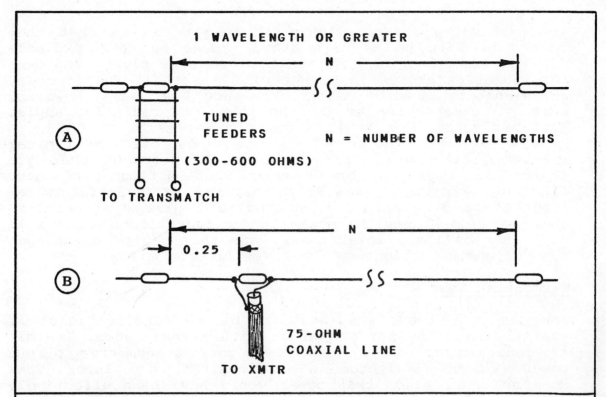

Fig 3-3 -- Suitable methods for feeding long-wire antennas. Feed-line balance will be best with the system at B. With method A the antenna suffers the same shortcomings of the end-fed Zepp in Fig 3-2. Balanced feeders may be used for the system at B. They should be 1/2 wavelength long, or multiples thereof. This will cause the low feed impedance to be repeated at the Transmatch or transmitter.

Long-Wire Characteristics

Perhaps you are wondering what advantages you can realize from using a long wire instead of a dipole. The primary benefit is that the system will provide gain over a dipole. The gain increases in proportion to the number of wavelengths used, assuming the antenna is high above ground. For example, the gain for a 3-wavelength long wire (referenced to a dipole) is 2 dB. A 6-wavelength wire yields a 4.8-dB gain, and a 10-wavelength long wire provides nearly 7-1/2 dB of gain. A gain curve that includes the radiation angles versus number of wavelengths may be found in chapter 7 of The ARRL Antenna Book, 14th edition.

The greater your long-wire length the higher the gain and the lower the radiation angle of the major lobes. This is ideal for DX work. Maximum radiation is off the ends of the wire, but numerous minor lobes exist at various angles, respective to the wire. I have found that there are lobes that are suitable for almost any type of propagation at a given period. Because of this feature, I have had, under certain band conditions, better results with a long wire than with my triband Yagi at 60 feet.

An interesting characteristic of long wires is that they do not need to be as high above ground as does a dipole for the same frequency, respective to directivity and useful radiation angles. A height of 30 feet is, for example, quite acceptable at 20 meters. You should be aware, however, that the greater the height the lower the radiation angle. Wave angles as low as 10 degrees are possible with large long wires, and angles of 15 to 20 degrees are typical when several wavelengths are used. I recommend that you design your long wire for 20 meters where optimum performance is most desirable, generally speaking. If several wavelengths are employed, the antenna will give good performance on 80, 40 and 30 meters. The radiation angles will be higher on these lower bands, and this can be advantageous for communications out to a few hundred miles.

Terminated Long Wires

You can cause your long wire to become unidirectional if you terminate the far end of it with a resistance. If this is done, approximately one half your transmitter output power will be dissipated in the terminating resistor. This does no harm, since that power would have been effectively wasted in the second major lobe (bidirectional) of your unterminated long wire. Fig 3-4 shows how to arrange the antenna for unidirectional radiation off the far end of the wire.

You may be wondering what the advantages might be when using a terminated long wire. If so, this is certainly a valid question. The notable advantage of this method is that QRM is greatly reduced during receive. The major antenna response is off the terminated end. Signals off the sides and back (transmitter end) of the wire are greatly reduced in strength. In other words, if your primary DX interest is associated with Europe, Africa or Oceana, for example, you can aim the wire at one of those regions and reduce QRM from other directions. Also, this technique may be useful in minimizing antenna response to prevailing storm fronts (QRN), such as those that prevail in the Gulf of Mexico. Response to some man-made noise sources may also be reduced or eliminated through proper orientation of a terminated long wire.

The terminating resistor for a long wire must be purely resistive. Wire-wound power resistors are not suitable because they are inductive. This presents an unwanted inductive reactance (XL) in the antenna system. Also, the resistor must be capable of safely dissipating one half the output power from your transmitter.

The terminating resistor needs to be connected to a ground screen at the far end of the antenna. A system of 1/4-wave in-ground or on-ground radial wires is best. You should also drive an 8-foot ground rod into the center of the radial circle where the radials converge. Connect the radial wires to the rod, then return the terminating resistor to the ground stake.

A substandard ground system will cause changing antenna performance with the seasons. Specifically, the moisture content in the ground affects the ground-system quality, and hence the general performance of the antenna. This is a common problem for those amateurs who rely on a single ground rod for the antenna termination point.

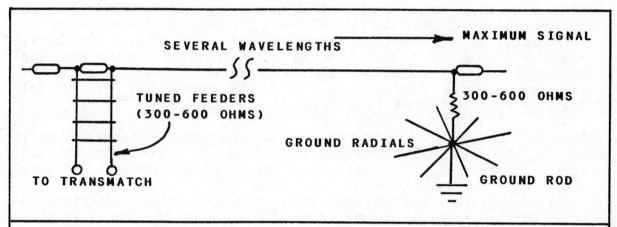

Fig 3-4 -- Arrangement for a terminated long wire. Quarter-wave radials and an 8-foot ground rod comprise the ground screen. Best performance will result when the antenna is a half wavelength or greater above ground. Use as many radials as you can, up to 120 units.

Chapter Summary

Substantially more technical information about the antennas in this chapter is presented in The ARRL Antenna Book. I recommend that reference as a source of data that will provide you with the theory of operation that I have avoided intentionally in this volume.

Single-wire antennas represent systems that are easy to build and erect. If you can accept the shortfalls of some of these antennas, they may be quite acceptable for your use.

Chapter 4

SIMPLE VERTICAL ANTENNAS

When we lack the space to "go out" with our antennas, we
generally have the option of "going up" with our radiators.
To go out with a wire suggests a horizontal radiator of
some type, whereas going up can be related to a large group
of vertical antennas.

What are the advantages and shortcomings of vertical antennas?
You will observe a vertical antenna to be noisier than a
horizontal one. Why is this? Well, most man-made noise is
vertically polarized (house wiring, drop lines and similar).
Maximum pickup of this noise will occur when our antenna
has vertical polarization. This QRN is not always a problem.
It depends upon our location and the number of electrical
devices that are used in our immediate area.

A vertical antenna, if it is a quarter-wavelength type,
or an odd multiple thereof, requires an effective ground
screen under it. This is seen as a disadvantage by some
amateurs. They don't want to spend the money required for
a ground-radial system, or they may lack the ambition to
bury numerous wires in the ground. So much for the undesir-
able features of a vertical antenna.

The good news about vertical antennas is that they require
very little real estate, and they produce a low radiation
angle. This is ideal for DX operation. A low radiation angle
is sought after among those of us who use horizontal antennas,
but it represents a minor tradeoff: a vertical that has
a low radiation angle is not as effective as a high-angle
horizontal radiator when communicating with stations that
are within a few hundred miles of us. But, a vertical will
usually provide a very strong signal within ground-wave
range (0 to 25 or 30 miles). Despite the seemingly negative
qualities of vertical antennas, they are much better than
no antenna at all!

Materials for Building Vertical Antennas

Perhaps you are thinking of massive, expensive structures
in connection with vertical antennas. You need not buy a
tower or steel mast in order to take advantage of vertical
antennas. Effective antennas may be made from wire, metal
downspout pipe, irrigation pipe, aluminum tubing or electrical-
conduit pipe. I once knew a ham who saved enough soup cans
to make a 40-meter vertical. He simply soldered the cans
together until he had 33 feet of antenna!

A base insulator may consist of a glass bottle or jug that
is set a few inches into the ground. Nylon guy ropes can
be used to keep the antenna erect. Bottles work nicely as
insulators when we use downspout material for the radiator.

Shortened Verticals

The rule that applies to short horizontal antennas holds
true for reduced-size verticals. When you trade physical
size for lower cost, the efficiency of the system must be
lower. There is no substitute for a full-size antenna, at
least in terms of optimum performance. Shortened antennas
may be resonated by means of loading coils and capacitance
hats, and they can be matched to 50-ohm feed lines. But,
the short antenna will have a narrower bandwidth, and there
will be losses in the loading coils. The amount of loss
in dB will depend upon the amount of shortening, plus the
actual losses in the loading coil. High-Q loading coils
with large-diameter wire are best. You should also be aware
that the higher the overall antenna Q the narrower the bandwidth of the radiator. You may consider the useful bandwidth
as that portion of a given ham band in which the SWR does
not exceed a 2:1 ratio. Because of these limitations, you
should try to make your vertical as long as is practicable.

The diameter of the antenna conductor affects the overall
system Q. The greater the cross-sectional area of the antenna
element the greater the bandwidth (lower Q). For example,
a quarter-wavelength 80-meter vertical made from no. 14
wire might have a 2:1 SWR bandwidth of 60 kHz. The same
antenna, if made from tower sections, may have a bandwidth
of 100 kHz. These are important matters to keep in mind
when you design your vertical antenna. The antenna bandwidth
may be improved by using several wires in parallel to form
the radiator. A conical monopole fits this description.

Where to Put the Loading Coil

You will hear controversial discussions about the best place
to install a loading coil for a short vertical or horizontal
antenna. Theory tells us that top loading is best. This
is because the current portion of an antenna radiates the
bulk of the RF energy. A top-loaded antenna does, therefore,
have the greatest area of current flow, compared to center-
or base-loaded antennas. Fig 4-1 shows the three methods
we may adopt when loading a short radiator. You will observe
that the center- and top-loaded versions include a "capacity
hat." This device lowers the resonant frequency of the system
by introducing additional capacitance across the coil. The
advantage is a smaller number of coil turns, and hence lower
coil loss. This procedure also increases the system bandwidth. The base-loaded system, on the other hand, uses the
driven element as the capacitance hat. Base loading requires
the fewest coil turns. Top loading dictates the greatest

number of turns in the inductor. A center-loading coil falls
somewhere between the top- and bottom-loading coils when
it comes to turns count. You may follow these same rules
when designing a horizontal antenna. The loading can be
thought of as feed-point loading, center loading or end
loading.

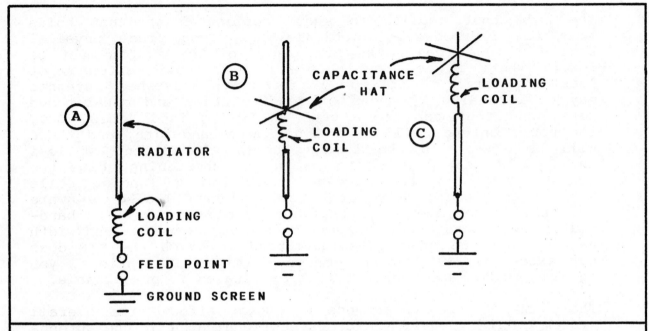

Fig 4-1 -- Illustration of a base-loaded (A), center-loaded (B)
and top-loaded (C) vertical antenna. Capacitance hats are
used with the loading coils at B and C to reduce the coil
inductance for a resonant system.

Loading-Coil Construction

Earlier in this chapter we learned that the Q of a loading
coil should be high (100 or greater) if we are to minimize
coil losses. This means that the coil wire should be of
heavy gauge, such as no. 14 or larger. The higher the RF
power applied to the antenna the greater the wire diameter.
If the wire gauge is too light, coil heating will result,
and this is wasteful of power. In other words, if the coil
runs warm when RF energy is fed to the antenna, power is
being dissipated in the loading coil.

The coil-form material should be of the low-loss type. An
ideal loading coil would be air wound, but this is impractical.
Ceramic or glass coil forms would be very good as loading-
coil foundations, but they break easily. What does this
leave as an alternative coil-form material? Solid phenolic
rod works well for coil forms. The fabric-base phenolic
is best for longevity, since it is less prone to moisture
absorption than is the paper-based phenolic rod. Another
insulating material that is worthy of consideration is a
plastic called Delrin. Lexan plastic is suitable also. These

materials are available from commercial plastics dealers, such as United States Plastic Corp. in Lima, OH 45801. This company accepts mail orders and publishes a catalog.

You should be aware that the form factor of coils has a bearing on the coil Q. A 1:1 form factor (length to diameter) yields the highest Q. For example, if the loading coil has a 3-inch diameter, the best Q will result when the winding is 3 inches long. Form factors of 2:1 (2 inch diameter and a 4-inch length, for example) are quite acceptable in terms of reasonable coil Q.

The coil Q can be aided by spacing the inductor turns. A spacing of one wire diameter between the turns is suggested. The tradeoff in doing this is a need for more coil turns for a given inductance. A close-spaced coil has greater distributed capacitance across it, and hence provides greater effective inductance than a coil with an equal number of spaced turns.

You should use enameled wire for close-spaced coils. The enamel insulation needs to be of high grade in order to prevent shorted turns. A shorted turn will destroy the coil Q and lower its inductance. Formvar enamel insulation is rugged, and it is relatively impervious to oil, acid and other air-born chemicals. No matter what type of wire you use, be sure to coat the coil (after it has been adjusted for resonance) with exterior polyurethane or spar varnish. This will help prevent deterioration of the enamel coating on the wire. You will provide greater protection against the effects of weather if you invert a plastic drinking glass or similar device and place it over the loading coil. This will keep sunlight off the coil (UV rays), and will prevent rain, snow and ice from detuning the coil.

If you have access to stranded wire with Teflon insulation, it will be to your advantage to wind your loading coils with this wire. Despite the high cost of Teflon wire, you will find the life span of your coil to be substantially greater than when using ordinary enameled wire. Teflon has a very high dielectric factor (good), and it is not affected by chemicals.

Do not discount the possibility of using high-Q Miniductor coil stock. This is available from B & W Corp., and the coil Q is high because it is essentially air-wound stock. It may be side-mounted on a Delrin or phenolic rod. This will ensure antenna strength while allowing you to use an air-wound loading coil.

Capacitance Hats

It may seem to you that we are putting the cart before the horse in this chapter. After all, we are addressing the subject of shortened vertical antennas before we discuss full-size ones. I feel that the topics we have covered thus far are germane to many types of vertical antenna, so let's continue along these lines for a while longer.

There is no special formula for determining the size of a capacitance hat. If I were to offer a rule of thumb for this device, I would say, "make it as large as you can." The greater the hat size the smaller the loading coil, and the lower the system losses. Fig 4-2 shows how to make a loading coil, along with suggestions for fabricating a top hat. If you have mechanical aptitude, I'm sure you can find a better way to engineer your loading coil and capacitance hat.

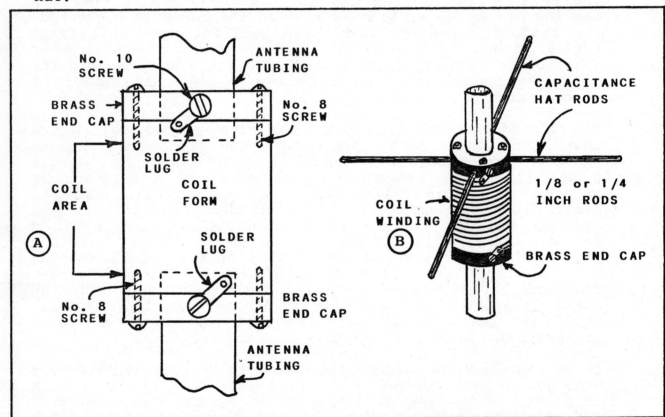

Fig 4-2 -- Mechanical details for a suggested loading-coil assembly. Four no. 8 machine screws hold each aluminum or brass end cap in place on the solid coil form. Two no. 10 machine screws are threaded into each cap to affix the coil form to the antenna rod or tubing. Drawing B shows how the capacitance-hat rods are attached to the brass end cap at the top of the coil form. The rods may be threaded to match tapped holes in the metal cap. Use the longest rods practicable, consistent with rigidity and resistance to wind. A metal disc may be substituted for the rods. More than four rods may be used at B to increase the effective capacitance of the hat.

You should consider the effects of dissimilar metals in your vertical-antenna system. For example, the loading coil in Fig 4-2 has brass end caps, but the antenna rod or tubing may be aluminum. This creates an opportunity for corrosion that will be worse than when you use like metals where an electrical joint exists. A solution to this problem lies in the use of aluminum tubing for the antenna, aluminum end caps for the loading coil and aluminum capacitance-hat rods. The attachment screws and two solder lugs will still be of a different metal than the remainder of the assembly. Electrolysis can occur at those points. I find that coating such parts with silicone grease helps to retard unwanted corrosion. Sealing these parts against moisture also helps. This may be done by placing noncorrosive silicone sealant over the heads of the screws. Caulking compound may be used for this purpose also.

Vertical-Antenna Impedance

Most quarter-wavelength verticals, when used over an effective ground screen, exhibit a feed impedance of approximately 25 ohms. Some of the short verticals may have feed-point impedances as low as 10 ohms, depending on the design. These antennas may be inefficient because of ground losses that can accompany an inferior ground screen. A large part of the RF power can be dissipated in the ground rather than being radiated, even though an impedance match is effected. Therefore, good antenna efficiency depends not only on the impedance match, but on a quality ground system.

Even if our vertical antenna exhibits a 25-ohm impedance, we still need to provide a match to 50-ohm coaxial line if we want maximum power transfer to the antenna. Maximum power transfer occurs only when an impedance match exists.

Various methods exist for matching a 50-ohm line to vertical antennas. We will get into this subject in considerable detail later in the book.

The notable exception to the above claim for a 25-ohm antenna impedance is when the vertical is mounted above ground, and it uses four or more above-ground radials as the counterpoise. If these radials are drooped toward ground at roughly 45 degrees, the feed impedance will approach 50 ohms. If you erect the same type of antenna, but have radials that extend straight out from the base of the vertical (90 degrees respective to the driven element), the feed impedance will be 25-30 ohms, and will require an impedance-matching device.

Popular Formats for Verticals

Let's focus our attention first on full-size quarter-wave verticals. You should strive toward a full-size antenna if finances and available materials permit this. Fig 4-3 has several pictorial representations of common vertical antennas.

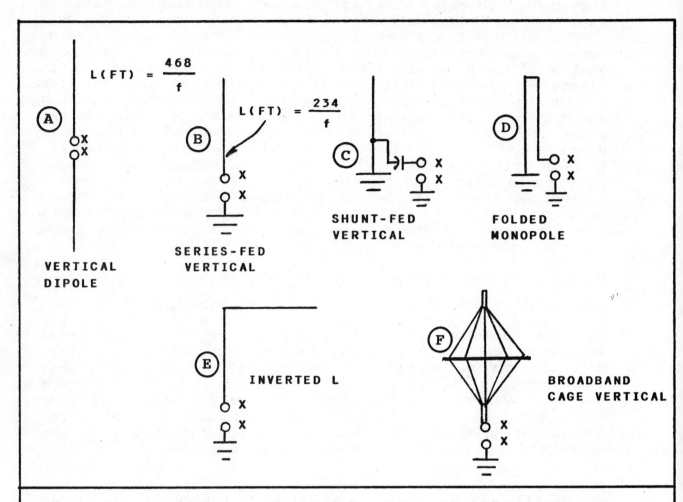

Fig 4-3 -- Various common forms for vertical antennas. A vertical dipole is depicted at A. The overall length is obtained from 468/f(MHz) = feet. Drawing B shows a series-fed 1/4-wave vertical. Antenna C is a grounded Marconi with shunt feed. A folded monopole 1/4-wave vertical is seen at D. It is equivalent to one half a folded dipole, and it has a higher feed impedance than the antenna at B. An inverted-L 1/4-wave vertical is illustrated at E. The bent top section acts as a top hat and does very little radiating. A broadband cage type of vertical is shown at F. Numerous wires are used in parallel to increase the bandwidth.

The vertical dipole of Fig 4-3A needs no ground screen, but it requires twice the height of the quarter-wave vertical antennas at B through F. This should not create serious mechanical problems at 14 MHz and higher. The series-fed verticals at B, E and F offer the greatest ease in matching. This is because the terminals are available for connection to matching networks. The shunt-feed method requires specific shunt-arm length, spacing from the radiator and shunt-arm diameter. The series capacitor in Fig 4-3C is made variable to aid the matching procedure. The shunt arm and capacitor form what is known as a <u>gamma</u> <u>match</u>.

The cage vertical of Fig 4-3F consists of a vertical element (tower or mast) to which is attached numerous wires that are fanned outward near the center of the radiator. This multiconductor antenna has a low Q, and hence greater bandwidth than single-conductor verticals. Although this style of antenna is somewhat unwieldy, you may wish to consider one for coverage of all of the 75/80- or 40-meter bands with an SWR of less than 2:1 across all of the frequency range. The larger the center circumference, and the greater the number of conductors, the greater the antenna bandwidth.

Above-Ground Verticals

The principles illustrated in Fig 4-3 may be applied to vertical antennas that are above ground, and which use "floating" radials as the ground screen. The advantage you will find when you are using above-ground vertical antennas is having the radiator elevated above conductive clutter, such as phone lines, power lines and metal fences. There will be less pattern distortion with an above-ground vertical that is high and in the clear. Also, there will be significantly less energy absorption with this kind of antenna, at least in comparison to a ground-mounted antenna that is surrounded by trees and conductive objects.

A secondary advantage associated with a vertical that is erected above ground is that you may use the radial wires as guy lines to hold the supporting mast erect. The radial wires need to have insulators at their lower ends in order to keep them isolated from guy-line extenders or earth ground. Refer to Fig 4-4 for information about the manner in which you may structure a vertical ground-plane antenna.

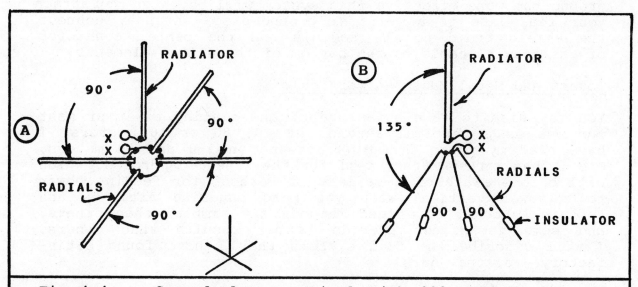

Fig 4-4 -- Ground-plane vertical with 90° radials (A) and a similar version (B) with drooping wire radials.

The antenna of Fig 4-4A is suitable for the upper end of
the HF spectrum (15, 12 and 10 meters), but represents a
mechanical obstacle at 14 MHz and lower. Aluminum tubing
is generally used for the elements of this antenna. The
four radial rods may be bolted or clamped to a metal plate.
The driven element is insulated from this plate by means
of a large standoff insulator or L bracket that has a piece
of thick-wall plastic tubing attached to it by means of
U bolts. PVC tubing is satisfactory as the base insulator,
owing to the low feed-point impedance. PVC tubing is not
recommended as insulation where high RF voltage is present.
The antenna driven element (vertical member) is cut to a
length that is determined by L(feet) = 234/f(MHz). Cut your
radials 5% longer than this dimension. The style of construction
shown in Fig 4-4A is ideal for VHF and UHF ground-plane
verticals. You may build a 2-meter version of this antenna
from coat-hanger wire or 1/8-inch diameter brazing rod.

You will observe a drooping ground-plane antenna in Fig
4-4B. You may fashion your radials from no. 12 or 14 copper
wire. As is the case with the Fig 4-4A antenna, the radials
are joined electrically at the center junction, but they
are isolated from the driven element. If you install your
radials so that they droop 45° (respective to the mast),
the feed impedance will provide a close match to 50-ohm
coaxial cable. You can experiment with the droop angle and
the length of the driven element to obtain an SWR of 1:1.

The antennas in Fig 4-4 will have a radiation pattern that
is omnidirectional. Vertical polarization will result from
this type of antenna, and the radiation angle will be low
(on the order of 15°). The greater the antenna height above
ground the more effective this system will be. When you attach
your feed line to a vertical ground-plane antenna, connect
the shield braid to the radials and the center conductor
of the coaxial cable to the bottom of the driven element.

Towers and Metal Masts as Verticals

You may wish to take advantage of an existing tower or mast
for use as a vertical antenna at 40, 80 or 160 meters. I
have used my tower for a DX antenna on 160 and 80 meters,
and it has worked rather well for the intended purpose. You
will become aware of a variety of methods for feeding power
to a tower vertical, when you read magazine articles and
technical books. Some feed methods are simpler than others,
and some techniques provide better results than others.
I will describe the feed systems that I have found satisfactory over the years.

There is no reason why you can't use your tower when it
has a triband Yagi and/or a VHF or UHF array atop it. These
additional antennas act as top-loading devices, and they
become a part of the overall radiator.

Fig 4-5 contains some examples of how to excite a tower as a grounded 1/4-wave Marconi antenna. The feed method you choose will be a matter of personal choice.

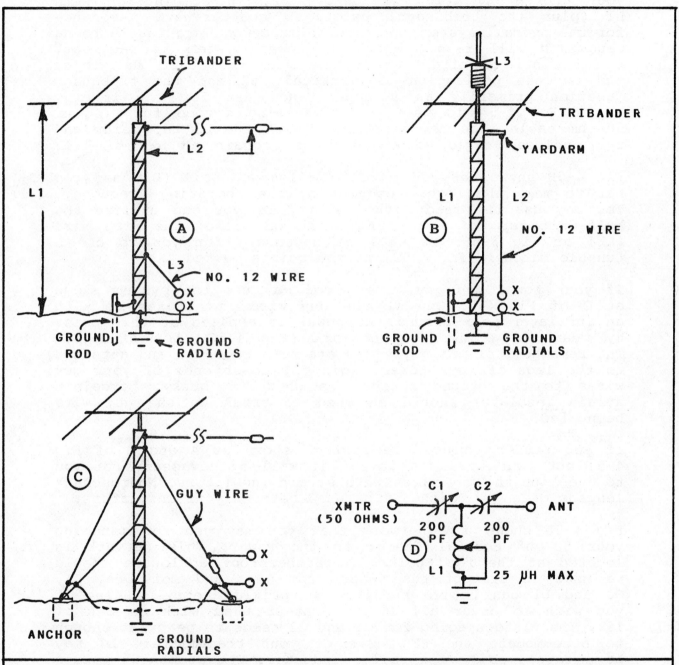

Fig 4-5 -- One half of a delta match is used at A to shunt feed a tower. A drop wire is used at B to provide a method of feed. Antenna C is a guyed tower with one guy wire used as a shunt-feed arm. The T network at D is suitable for use at the antenna feed point. It may be used to obtain an SWR of 1:1. L networks are suitable for this purpose also. See the text for further discussion about these antennas.

You will notice a horizontal extender wire (L2) in the examples of Fig 4-5A and C. This wire is used when the combined tower and Yagi antenna provide resonance above your chosen operating frequency. L2, in combination with L1 (plus the tribander) establish quarter-wave resonance for the overall system. Best performance and greatest antenna bandwidth will result when the tower system is resonant. L2 will not be effective as the main radiator. Therefore, the antenna polization is vertical. L2 and the tribander function primarily as a capacitance hat for the tower. L3 of Fig 4-5A is used as a matching device for your 50-ohm cable. You may experiment with the L3 tap point and the enclosed angle of the wire to obtain an SWR of 1:1.

Fig 4-5B shows how you may use a loading coil (L3) and top hat to make the tower resonant at the operating frequency. You may use the feed method shown, or you can utilize the half-delta arm of Fig 4-5A. When you employ the drop wire (L2) of Fig 4-5B, it will be necessary for you to use a tunable matcher (Fig 4-5C) at the points marked "X."

If your tower has guy wires, you may use the system shown at C of Fig 4-5. One of the guy wires is broken up with an insulator near ground. RF power is applied at this point by means of a matching network (Fig 4-5C, for example). The remainder of the guy wires are left grounded and attached to the legs of your tower. Join the lower ends of your guy wires to the ground screen, as shown. A husky, porcelain strain insulator should be used to break up the guy wire being fed.

If you prefer to use the systems shown at A and B of Fig 4-5, but your tower is guyed, it will be necessary for you to break up the guy wires with strain insulators. Nonresonant lengths of guy wire should be used between your insulators.

Fig 4-5D shows a T network that you may use for matching your 50-ohm coaxial feeder to the tower. This network is located at the feed point. A weatherproof enclosure should be used to protect the tuner from dirt and moisture. C1, C2 and L1 can be motor-driven to provide remote tuning if you wish to cover all of an amateur band with an SWR of 1:1. The plate spacing for C1 and C2 needs to be great enough to accommodate the RF power of your transmitter. L1 may be a tapped coil or a roller inductor. The component values shown in the diagram are suitable for 160- and 80-meter operation. Smaller values of C and L are best for 40-meter operation.

If your tower is too tall for resonance in your chosen band of operation, use the system illustrated at B of Fig 4-5. You will need to eliminate the top-loading coil, L3.

The electrical integrity of your tower joints is essential to proper antenna performance. If you aren't sure of how good your tower joints are (oxidation), place a jumper across each section on one of the legs. The shield braid from RG-8 coaxial cable works nicely for this purpose. You may find it necessary also to place a shield-braid loop between the top of your tower and the mast that holds your triband Yagi. A poor electrical connection at that point in the system will cause SWR variations and noise during periods of wind.

Rotator Cables and Feed Lines

You are probably wondering what effect the rotator cable and tribander feed line will have on your shunt-fed tower. I have been asked this question countless times when giving an antenna lecture. If these cables are taped to the tower legs and brought all of the way to ground, no problems will be encountered. When you bring these wires to ground, they are effectively at zero RF potential. It is wise to bury the control cable and feed line enroute to your house. This will help to prevent unwanted RF energy from being picked up by these conductors after they leave the antenna. The foregoing treatment of feed lines and other cables ensures that they actually become a part of the shunt-fed tower, but do not cause RF to enter the shack by some devious route. I have never experienced problems with my rotator control box (lights coming on when they shouldn't, etc.) when shunt-feeding a tower with an HF beam antenna atop it.

Loading Effect of Yagis

Perhaps you're curious about the effect your tribander may have on overall antenna resonance when you shunt feed your tower. I made some tests along these lines when I shunt-fed a 50-foot Rohn-25 tower (unguyed). I had a Cushcraft A-4 tribander on my tower during these tests. With the system resonant at 1.9 MHz (extender wire used), I removed the Yagi. System resonance moved upward to 2.0 MHz. The basic tower and Yagi (not arranged for 160-meter use) showed a resonance at 4.2 MHz when I checked it with a dip meter. You may do this by adding a temporary drop wire (Fig 4-5B), then placing a small 6-turn coil between the lower end of the drop wire and ground. Insert the dip-meter probe into this coil and learn the resonant frequency. The mast that supported my tribander was five feet long.

SWR Bandwidth of Towers

You will find the bandwidth greater with the system of Fig 4-5A and C than is the situation with the top-loaded vertical of Fig 4-5B. Also, the extender-wire method offers slightly greater antenna efficiency (coil losses). The antennas at A and C of Fig 4-5 should yield a 2:1 SWR bandwidth of 100 kHz on 160 meters. Conversely, a nonresonant short tower may provide a 2:1 SWR bandwidth of only 10 or 15 kHz at

1.9 MHz. In this example you should assume that you are using a basic tower and tribander, and minus a top-loading coil or extender wire.

The antenna bandwidth on 80 meters should be approximately twice that for 160 meters, assuming that you have structured it for 80-meter operation. This is because the bandwidth of a resonant circuit doubles at the next harmonic, provided the Q of the system remains the same. Therefore, if your tower exhibits a 100-kHz bandwidth at 1.9 MHz, the same tower (when resonant) will provide a 200-kHz bandwidth at 3.8 MHz, and so on.

<u>The Classic Gamma Match</u>

Although our discussion of matching devices will be presented later in this book, I want to illustrate how the dimensions are derived while we are still considering shunt-fed towers or masts. Fig 4-6 provides approximate dimensions, along with the factors you will use to arrive at the correct sizes for the gamma arm, spacing from the tower and the ball-park capacitance for the reactance capacitor.

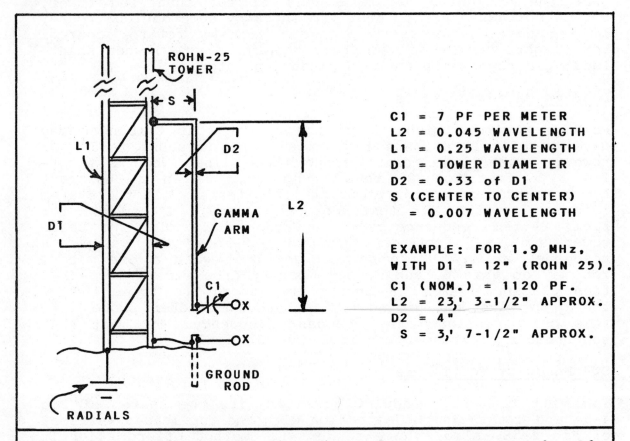

C1 = 7 PF PER METER
L2 = 0.045 WAVELENGTH
L1 = 0.25 WAVELENGTH
D1 = TOWER DIAMETER
D2 = 0.33 of D1
S (CENTER TO CENTER)
= 0.007 WAVELENGTH

EXAMPLE: FOR 1.9 MHz,
WITH D1 = 12" (ROHN 25).
C1 (NOM.) = 1120 PF.
L2 = 23,' 3-1/2" APPROX.
D2 = 4"
S = 3,' 7-1/2" APPROX.

Fig 4-6 -- Approximate dimensions for a gamma match. L1 should be shortened approximately 3% from the resonant height when using gamma matching. Performance is dependent upon a quality earth screen.

You can think of a gamma-matching section as one half of a T-matching section, which is sometimes used to provide balanced feed to a driven element that is not split at its electrical center. The approximations listed in the drawing of Fig 4-6 will enable you to arrive at workable dimensions when feeding your tower by means of a gamma section. A more scientific design procedure is given in **The ARRL Antenna Book,** chapter 2.

The gamma arm of Fig 4-6 requires a fairly large diameter in the example shown (4 inches). This is because of the large cross-sectional area of the tower, since D1 and D2 are directly related. If you use a telescoping mast or some other small-diameter radiator, the D2 problem becomes easy to solve. I have built a 4-inch diameter gamma arm by using several lengths of no. 14 stranded copper wire in a cage configuration. I used pieces of circuit-board material (single or double sided) as the spacers that held the wires apart. The PC-board pieces were 5 inches in diameter. I spaced them three feet apart along the span of the gamma arm. The wires (6 each) passed through holes around the perimeter of the spacers, and were soldered in place at those points. This type of construction seems to provide results that are equal to those obtained from a solid 4-inch diameter conductor. It is possible that a suitable length of metal downspout or small-diameter stove pipe can be used as a gamma arm. If this is done you will need to pin the joints with sheet-metal screws, then solder the seams of the joints.

The information in Fig 4-6 is based on an antenna that is a resonant quarter wavelength. If your tower is too short to meet this requirement, use the extender wire or loading coil shown in Fig 4-5. Also, it is essential that you make the overall antenna slightly shorter than 1/4 wavelength when using a gamma match. For example, if you choose 1.9 MHz as the center frequency for your antenna, it will be resonant at 1956 kHz for proper gamma-match operation. This equates to a 3% upward shift in frequency. You may use your dip meter and a small sampling coil at the bottom of the gamma arm to check the resonant frequency of the system. C1 is not used during this test. The sampling or coupling coil may consist of 6 or 8 turns, 1 inch in diameter. Place it between the lower end of the gamma arm and the system ground.

C1 of Fig 4-6 is adjusted to obtain an SWR of 1:1 when using a 50-ohm feed line. You may want to measure the capacitance of C1 (1:1 SWR), then replace C1 with a fixed-value capacitor of the same value. I once made a fixed-value capacitor from 3-foot sections of telescoping aluminum tubing. The outer tube had an OD of 1.5 inches. A 1.25-inch OD tubing section served as the plunger for varying the capacitance. The inner tubing section was wrapped with polyethylene sheeting (four layers), then affixed in four places with Scotch tape. This served as the dielectric material for the homemade tubular capacitor. I adjusted the device to obtain an SWR of 1:1, then I taped the tubing sections in position. I mounted the capacitor on the base of the tower (on insulating blocks) and sealed each end to keep out the dirt and moisture. It served nicely as a high-power, fixed-value capacitor.

If you use an air variable capacitor at C1 of Fig 4-6, make certain the plate spacing is suitable for the RF power from your transmitter. You may want to use a 1-RPM electric motor to adjust C1 from within your shack. This will enable you to keep the SWR low when you QSY. C1 and the motor can be housed in a weatherproof enclosure.

The suggested dimensions in Fig 4-6 are based on the free-space wavelength formula [L(feet) = 984/f(MHz)]. The overall radiator length is, however, determined from the standard equation: L(feet) = 234/f(MHz). If your tower has a beam antenna atop it, the calculated radiator dimension will need to be shortened somewhat to compensate for the loading effect of the beam antenna.

The 5/8-Wavelength Vertical

It is practical to use a 5/8-wave vertical from the upper end of the HF spectrum into the UHF region. This antenna offers some gain over a dipole or quarter-wave vertical (approximately 1.8 dBD in practice). It has a somewhat higher radiation angle than a quarter- or half-wavelength vertical antenna, which can be advantageous when we use this antenna for repeater access from a vehicle. Typically, the VHF or UHF repeater is located at some high point in the area. Because of this, the 5/8-wave antenna can deliver a stronger signal at the repeater antenna than would be the case with a 1/4-wave vertical with its lower radiation angle. This larger mobile antenna reduces flutter or "picket-fencing" when used to operate through repeaters.

Although a 5/8-wave antenna is referred to as such, it is an electrical 3/4 wavelength, hence the higher radiation angle, compared to a half-wave dipole, for example. The coil at the base of the antenna is equivalent to 1/8 wave-length, electrically. This, plus the 5/8-wave whip, makes the overall antenna a 3/4-wavelength device. The 3/4-wave electrical property provides a close match to a 50-ohm feed line.

By itself, a 5/8-wavelength conductor presents an Xc (a capacitive reactance) of approximately 165 ohms. The series base coil cancels this reactance to make the antenna resist-ive at the operating frequency. In other words, the coil XL must also be 165 ohms.

Our knowledge of the Xc for the 5/8-wave antenna permits
us to calculate the required series inductance for cancelling
the reactance of the antenna: L(μH) = X/[f(MHz) X 6.28)]. We
will use as an example a 5/8-wave antenna for 28.6 MHz:
L(μH) = 165/(28.6 X 6.26). Our coil must have an inductance
of 0.92 μH. This is not an absolute value, owing to the
dielectric effects of the coil form and stray capacitance.
You may find it necessary to vary the coil inductance and
the length of the radiator in order to obtain an SWR of
1:1 for a 50-ohm feed line.

Fig 4-7 shows various forms you may adopt when building
a 5/8-wave vertical antenna. You will see examples of series
and shunt feed in the illustration.

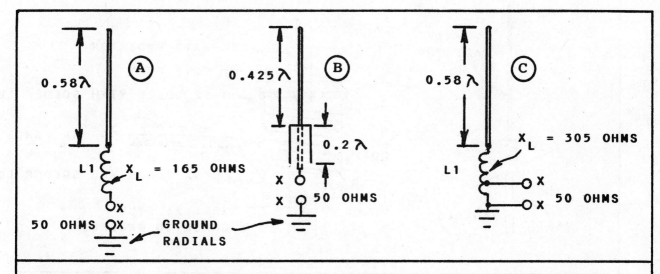

Fig 4-7 -- Examples of 5/8-wavelength verticals. Series
feed is seen at A. Antenna B needs no coil, but uses a 0.2-
wavelength sleeve at the feed point. The system at C uses
shunt feed. The coil tap is adjusted to obtain an SWR of
1:1.

Refer to Fig 4-7. You will notice that L1 at A has a lower
reactance than L1 at C. This is necessary in order to provide
a tap point at C, whereas no tap point is required at A.
Adjustment of the antenna of Fig 4-7A is done by varying
the length of the radiator and/or adjusting the turns of
L1. Tune the system for an SWR or 1:1. The shunt-fed antenna
at C of Fig 4-7 requires adjustment of the radiator and
the L1 tap to obtain an SWR of 1:1. The sleeve antenna in
Fig 4-7B has no coil. The radiator length is varied to obtain
a low SWR. The 0.2-wavelength sleeve should have a 2:1 or
greater ratio, respective to the OD of the radiator. In
other words, if the radiator has a 0.5-inch OD, the sleeve
should have a 1-inch or greater OD. All of these antennas
require above-ground 0.28-wavelength radials, or a large
solid metal plane (such as an auto body) under them. HF
versions of the 5/8-wavelength vertical may use buried ground
radials.

It is practical to construct 5/8-wave verticals as low in frequency as 7 MHz. If you do not have a support of adequate height for these longer antennas, you may slope them from a tower, mast or tall tree. I once used a 40-meter 5/8-wave antenna that sloped at a 45°angle. The performance was good, and the polarization remained vertical. I used the shunt-feed method shown at Fig 4-7C.

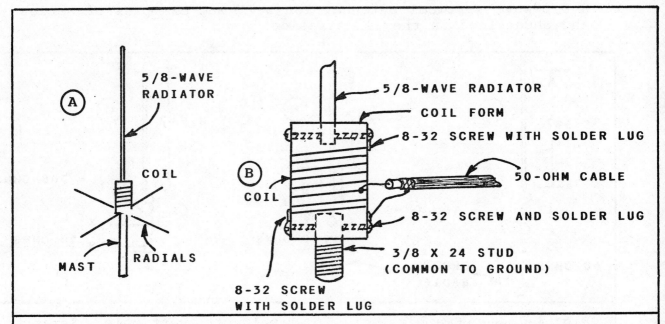

Fig 4-8 -- A mast-mount type of 5/8-wave vertical is shown at A. The radials are 0.28-wavelength long. A second set of radials are sometimes mounted 0.5 wavelength below the top set to decouple the mast from the antenna. This improves the radiation pattern. The drawing at B illustrates a method for building a 5/8-wave vertical for mobile use. The coil is for a shunt-fed system (Fig 4-6C).

When using the antenna of Fig 4-8A, route your 50-ohm feed line through a hole in the mast, just below the radials. The feeder is kept inside the mast up to the takeoff point to the station.

You may use phenolic or Delrin rod for the coil form (Fig 4-8B). The 8-32 screws hold the whip and the stud in place while making the necessary electrical contacts at each end of the coil. For operation at 144 MHz and higher, use a 1/8-inch OD stainless-steel rod for the whip. A telescoping auto-radio antenna may be substituted for the steel rod.

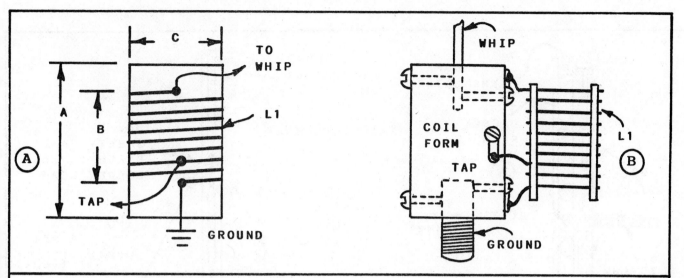

Fig 4-9 -- The drawing at A has designators that relate to Table 4-1. The 5/8-wave antenna coil assembly at B is suggested for use at 28 and 50 MHz, where the radiator is long and heavy. The coil form may be solid phenolic rod or a plastic rod that is of high dielectric quality.

Table 4-1 lists coil-winding data along with some suggested dimensions for L1 of Fig 4-9. You may use coil diameter and winding lengths that differ from those listed, provided the coil inductance is the same.

TABLE 4-1					
FREQ. (MHz)	L1	TAP	A	B	C
28.5	1.7 µH. 15 ts no. 10 wire.	4 ts	4 in.	3 in.	1 in.
51.0	1.0 µH. 8 ts no. 12	2 ts.	3 in.	2 in.	1-1/4 in.
146.0	0.33 µH. 7 ts. no. 14	1-3/4 ts.	2-1/2 in.	1-1/4 in.	3/4 in.
221.0	0.22 µH. 5 ts. no. 14	1-1/4 ts.	2-1/2 in	1-1/4 in.	3/4 in.

The number of coil turns and tap points may vary with the type of coil form used. The above data is for an air-wound coil. The dielectric factor of the coil form may increase the effective coil inductance. This will require a slight reduction in the coil size. For each listing above, the radiator is 5/8-wave long. For 28.5 MHz, 20 feet; for 51 MHz, 11 feet, 2-1/4 in.; for 146 MHz, 47 inches; for 221 MHz, 2 feet, 7 inches. Adjustment of the radiator length will also affect the SWR. Completed coils should be protected from dirt and moisture by covering them with tubing that has a high dielectric factor, such as polystyrene of Plexiglass.

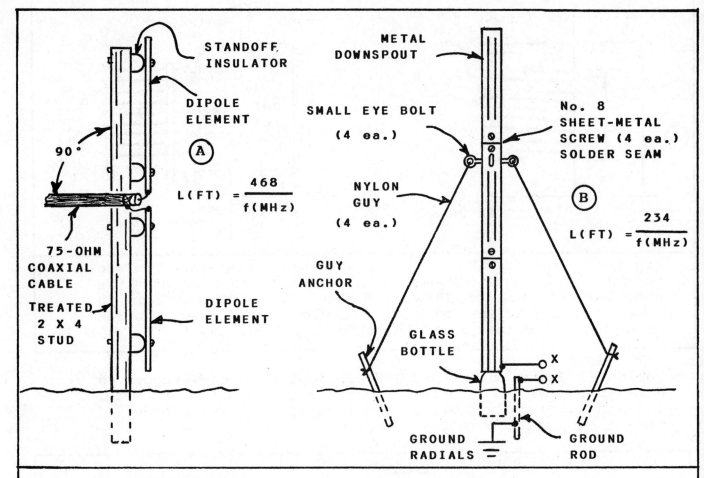

Fig 4-10 -- Illustrations of practical verticals. Drawing A shows how to erect a simple, inexpensive half-wave dipole on a wooden support pole. The 1/4-wavelength vertical at B may be constructed from metal downspout material. A glass bottle or jug serves as the base insulator.

Keeping the Cost Low

Before we wrap up this part of chapter 4, I want to encourage you to build your own vertical antennas. Not only will you benefit from the "learn by doing" activity, you will save a substantial amount of money in the process. Ham-radio ingenuity will reveal a number of ways to "make do" with ordinary materials. Junk yards offer a variety of materials that can be used for antenna conductors. You will discover such treasures as copper water pipe, electrical conduit and old electrical wiring in scrap yards. Used lumber may also be purchased at low prices, and you may use the material for your antenna supports.

Fig 4-10 shows two examples of low-cost vertical antennas. Mount the vertical dipole as high above ground as you can. This will increase the desirable low-angle radiation. Try to bring the feed line away from the dipole at a right angle for at least one half wavelength. This will help prevent the cable from disrupting the radiation pattern.

The vertical antenna in Fig 4-10B is made from low-cost gutter downspout material. The joints need to be pinned with sheet-metal screws, then soldered to provide a good electrical bond. A propane torch is useful for this part of the job. Small eye bolts serve as tie points for the nonconductive guy lines. You may use metal guy wire if you break the guy lines into nonresonant lengths with insulators. A gallon jug or similar large glass container is suitable as a base insulator for the antenna. Downspout verticals are entirely practical from 10 to 30 meters. The taller verticals may require two sets of guy lines.

If you want to erect, for example, a vertical ground-plane antenna to use on 20 meters, but wish to avoid the high cost of aluminum tubing and a metal support mast, consider the arrangement in Fig 4-11. The entire syetem is composed of no. 12 wire and six insulators. The antenna is supported by a tree limb or a nylon rope that is strung between two available supports. I have used this technique when camping and operating Field Day. The antenna performed as well as my all-metal version at home!

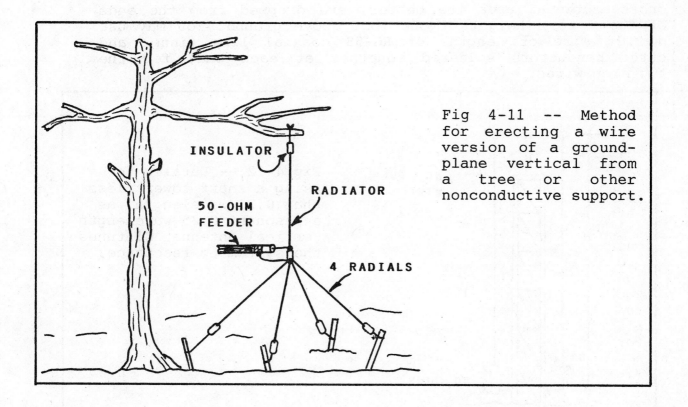

Fig 4-11 -- Method for erecting a wire version of a ground-plane vertical from a tree or other nonconductive support.

Physically Short Verticals

There are times when, for aesthetic or practical reasons,
we do not wish to have a tall tower or mast on our property.
Our logical option is to erect a short vertical antenna.
It can still be the proper electrical length if it is a
resonant 0.25 or 0.5 wavelength. Various means are available to us for establishing antenna resonance. The important
thing to remember when dealing with shortened antennas is
that we must always trade performance for size. The antenna
efficiency will not be equivalent to that of a full-size
radiator. However, in practice we may not recognize this
degraded performance if we do not have a full-size reference
antenna to use as a comparative radiator. Also, the loss
in dB for the smaller antenna may be only 1 to 3 dB, which
is difficult to discern by ear.

Short Three-Conductor Monopole

Fig 4-12 shows the details for using a tower or mast that
is less than 1/4 wavelength high. You may adopt this system,
even if your tower has a triband Yagi atop it, provided
the overall tower and beam antenna is less than an electrical
1/4 wavelength.

A T-shaped structure is placed at the top of the tower as
shown in Fig 4-12. Two large conductors (the larger the
cross-sectional area the better) are dropped from the ends
of the T to a height of one foot above ground. You may use
no. 10 wire or lengths of RG-58 coaxial line (inner and
outer conductors soldered together at each end) for the
two drop wires.

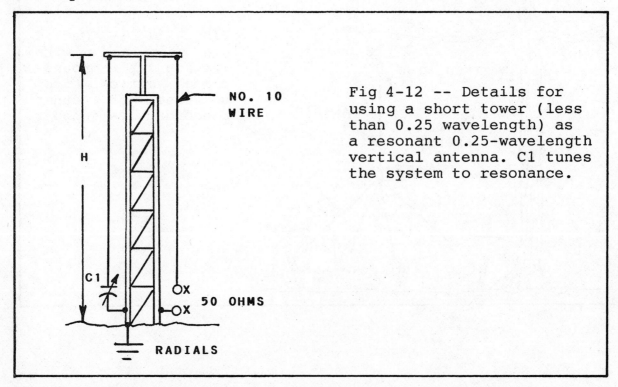

NO. 10 WIRE

H

C1

50 OHMS

RADIALS

Fig 4-12 -- Details for
using a short tower (less
than 0.25 wavelength) as
a resonant 0.25-wavelength
vertical antenna. C1 tunes
the system to resonance.

The antenna in Fig 4-12 suffers from narrow bandwidth, as is the case with most short verticals. I used a 35-foot tower as a 75-meter vertical when I tested this system. The 2:1 SWR bandwidth was 20 kHz. A remote-controlled tuning capacitor (C1) may be used to enable you to QSY within the band of interest. C1 must have wide plate spacing to prevent arcing when RF power is fed to the antenna. Very high values of RF voltage are present at C1. My test model had a feed impedance very close to 50 ohms. This can vary with the tower height and the spacing between the tower and drop wires. I used a spacing of one foot. An impedance-matching network may be required at the antenna feed point. Antenna performance was entirely acceptable for DX work on 75 meters.

Top-Loaded Vertical

You will see a short, top-loaded vertical antenna in Fig 4-13. The loading coil is at the top of the radiator. Some rods are used as a capacitance hat.

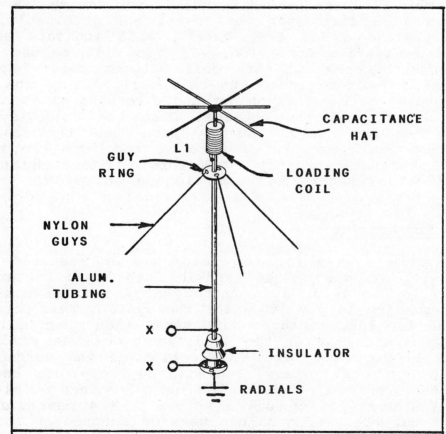

Fig 4-13 -- Short, top-loaded vertical. The capacitance hat increases the band-width and improves the efficiency. Nylon guy lines are used.

This antenna may be any height less than 0.25 wavelength. However, the shorter the antenna the larger the loading coil, and the lower the overall efficiency. In other words, you could actually use this technique to construct a 6-foot antenna for 160 meters. Try to make the antenna as tall as you can. The improved performance will reward you.

The greater the capacitance-hat size the better the performance. This is because fewer turns of wire will then be needed for the loading coil. Fewer turns mean lower ac resistance. A too-large capacitance hat will make the system awkward, and it may present mechanical problems.

Nylon guy lines are used for the antenna in Fig 4-13. If you use metal guy lines, they must be insulated from the vertical element of the antenna. You may use a glass bottle for the base insulator.

You can use a dip meter or SWR indicator to determine the proper number of loading-coil turns. If you use an SWR bridge, adjust the coil turns for the lowest SWR obtainable. This may not provide a 1:1 SWR, but it will indicate antenna resonance (resistive condition). If you plan to use a dip meter, insert a small 5-turn coil between points X, then insert the dip-meter probe into this coil. Adjust the turns on the loading coil until you get a dip reading at the desired center frequency of the antenna. Once the coil is close to being "on the nose," you may fine tune the resonance by trimming small amounts of wire or rod from the tip of one of the capacitance-hat conductors. If the antenna is too high in frequency, add some length to one of the hat wires.

Improved Top Loading

I developed a system for top loading that resulted in an antenna with efficiency and bandwidth that was better than the model in Fig 4-13. A large top hat was made from wires that served also as guy lines for the system. This permitted me to use far fewer loading-coil turns than when using the small capacitance hat in Fig 4-13. My model was structured as shown in Fig 4-14. The top section of the vertical is 10 feet high for 80 meters. I used a 20-foot lower section. The two radiator pipes were made from 1-1/2 inch OD aluminum tubing. You may want to use a wooden 2 X 4 mast for your model. If so, the conductor can be side mounted on the mast by using standoff insulators. This will permit you to use no. 10 wire, copper tubing, or electrical conduit for the two vertical sections of the antenna. It will also eliminate the need for a base insulator.

If you use the antenna as it is shown in Fig 4-14, be sure to use insulated guy lines below the top-hat section. This will prevent the lower part of the guying system from affecting the antenna performance.

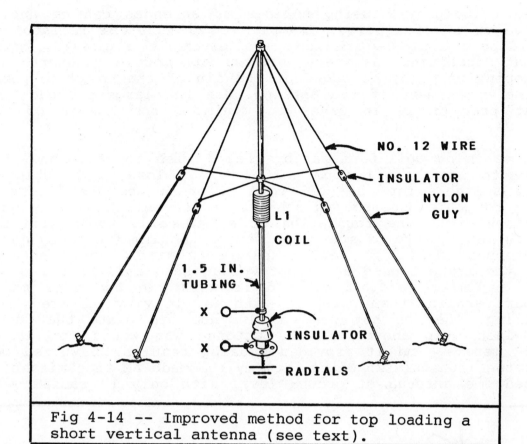

Fig 4-14 -- Improved method for top loading a short vertical antenna (see text).

Helically Wound Short Verticals

Good performance may be had when you use a long helix as a radiator. This requires a low-loss coil form that is fairly long. A helically wound vertical resembles a very long coil, since it is wound with wire from one end to the other. A common example of a helically wound vertical is the "rubber duckie" antenna used on VHF hand-held transceivers. The antenna is an electrical quarter wavelength, but it is a great deal shorter, physically, than a full-size counterpart. You will find that the efficiency is on par with most other short antennas -- lower than that of a full-size 1/4-wavelength radiator.

A capacitance hat is necessary for the helical antenna. Not only does the hat aid efficiency and bandwidth, it helps to prevent the tip of the vertical from arcing and burning. A helical antenna will, when fed sufficient power, develop very high RF voltage at the top end (somewhat like a Tesla coil). I once wound a 75-meter helical antenna on a 5-foot piece of dowel rod. It had no capacitance hat, and it caught fire at the far end with only 50 watts of RF power applied to it!. Addition of a 10-inch whip cured the problem. A metal extension draws the high voltage away from the helix and prevents the corona effect.

The principle of using helices as antennas can be applied also to short dipoles. Examples of a vertical radiator and a dipole that use helices are given in Fig 4-15. Notice that capacitance hats are used in both models. A short tele-scoping whip may be added at the tip of the helix to permit fine adjustment of the antenna. Use the largest capacitance hat practicable in order to minimize the number of coil turns.

There is no coil-form factor that I wish to recommend. But, try to use a coil form that has low loss. PVC tubing is not suitable for the coil form. I prefer to use fiberglass rod for this purpose. An athletic pole-vaulting rod is quite suitable for the foundation of a helically wound vertical. This is the type of material used for medium-frequency marine antennas. During my early experiments with helical antennas in the 1950s I used a 16-foot wooden hand rail that I bought at a lumber yard. I covered it with two coats of marine spar varnish, then wound it with no. 14 vinyl-covered house wiring. The top hat was an aluminum pie plate that had a 20-inch whip above it. The antenna was wound for use on 160 meters, and it provided amazing results. I worked many states with only 25 watts from my homemade AM transmitter. I used the antenna at ground level with only 10 radial wires.

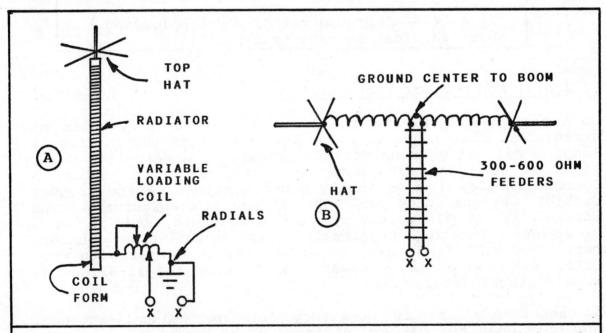

Fig 4-15 -- Example of a helically wound vertical (A) and a helically wound dipole (B) that uses tuned feeders.

Try to have a space equal to one wire diameter between the turns of the helix. This may be done by using a nylon rope as a spacer during the winding process. The rope and the wire are laid on the coil form at the same time. Apply a fair amount of tension to the wire and rope as you wind your helix. This will ensure that the wire winding will remain in place when you unwrap the nylon rope. The coil turns may then be affixed by painting the helix with exterior polyurethane or marine spar varnish. Two coats are best.

You may use enameled wire, close wound, for your helix. It is better, however, to have some space between the turns in order to prevent shorted coil turns later on. Use the lowest wire gauge possible, consistent with the required number of coil turns. The greater the cross-sectional area of the wire the lower the losses.

As for the necessary number of helix turns, I can only offer a rule of thumb that has worked for me. Wind your helix with a half wavelength of wire for the operating frequency. I have learned that it takes slightly less than one half wavelength of wire to obtain a resonant quarter-wave helical antenna. There will be some extra turns when you complete the winding. These will give you leeway when you prune the coil to resonance (after the capacitance hat is in place atop the helix).

A dip meter or SWR bridge can be used to determine the system resonance, as was explained during the discussion of Fig 4-13. It is necessary to have your ground radials in place before you adjust the coil length.

One advantage of a helically wound vertical over one that has lumped inductance (bottom, center or top loading) is that the voltage and current distribution along the radiator is uniform. This is not the situation with other types of loading (Fig 4-13, for example).

A variable inductor is shown in Fig 4-15A. If you make your helix resonant above the desired operating range, you may use the variable matching coil to resonate the antenna across all of a given amateur band. Frequent readjustment will be necessary. The coil tap provides a match to 50-ohm feed line. A piece of large Miniductor stock is handy as the matching coil. You may add taps for various parts of the band, then use a switch to select the coil taps.

The helically wound dipole of Fig 4-15B can be fed with coaxial cable by linking into the center of the dipole. Experiment with the number of link turns to obtain a 1:1 SWR. Typical bandwidth with this type of dipole at 7 MHz is approximately 50 kHz between the 2:1 SWR limits. Open-wire or molded balanced feed line can be used by tapping the feeder conductors out from the center of the helix, as shown. I tapped my 40-meter helix 10 turns each side of center (not critical). Tuned feeders will enable you to cover all of an amateur band by means of a Transmatch.

160-Meter Broadband Wire Vertical

The principle illustrated in Fig 4-14 may be applied when using a short vertical antenna made entirely from wire. I developed the antenna of Fig 4-16 for use on 160 meters. This antenna is described in detail in **QST** for Nov. 1986, page 26.

The top wire sections comprise a 6-foot top hat for loading coil L1. The lower wire section has two conductors to broaden the antenna response (greater conductor thickness). The dashed lines in Fig 4-16 represent a third wire you may add to broaden the response further. The top of L1 connects to the capacitance hat, and the bottom of the coil is common to the lower wire section.

My antenna was 60 feet long, overall. Therefore, the bottom portion was 54 feet long. Thin-wall, high-impact polystyrene tubing (3/4 inch OD) may be used as lighweight spreaders for the wires. I bought my tubing from United States Plastic Corp. in Lima, OH. You may use 1/2-inch PVC tubing if you wish, but it will greatly increase the antenna weight. Dowel rod (wooden) may be used for spreaders if it is treated first with exterior polyurethane varnish, or if you boil the spreaders in canning wax.

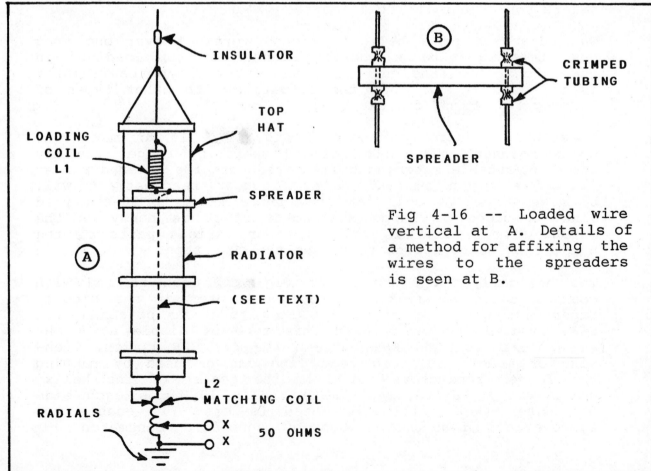

Fig 4-16 -- Loaded wire vertical at A. Details of a method for affixing the wires to the spreaders is seen at B.

L1 of Fig 4-16 has only enough inductance to cause resonance just outside the upper limit of the operating band. My 160-meter wire vertical was resonant at 2.8 MHz with only L1 in the circuit. L2, the matching/loading coil, makes up the difference to provide resonance. All of the coil pruning is done at L2. After the system is made resonant at the center of the chosen operating range, you may experiment with the tap point on L2 to obtain a 1:1 SWR. As you change the tap point you may need to readjust L2, overall, in order to keep the system in resonance. All tests must be made with the antenna referenced to your radial system. Do not attach the feed line to the L2 coil tap until the antenna resonance is established. Otherwise, you will get false readings on the dip meter.

Fig 4-16B shows how to lock the spreaders in place on the wires. You may use short pieces of 1/4-inch copper tubing. Crimp them with a crimping tool or diagonal cutters until they are tight on the wires. I used some thin-wall aluminum spacers for this purpose. This method is useful when building homemade open-wire feeders.

The 2:1 SWR bandwidth of my 60-foot, 160-meter antenna of Fig 4-16 was approximately 165 kHz while using no. 14 copper wire (stranded) for the conductors. The third wire (dashed lines) should increase the bandwidth another 10 or 15 kHz. My system had 20 on-ground radials, each 1/4 wavelength long. The antenna sloped from a 50-foot tower, but still worked well and produced vertically polarized waves. Numerous tests indicated it to be on par with, or sometimes 3 dB worse than my full-size 160-meter inverted L at 50 feet.

You may adopt this design for 75 or 40 meters by scaling it down. The 75-meter bandwidth should be roughly twice that for 160 meters, and the 40-meter bandwidth will be approximately twice that of the 75-meter version.

Trap Verticals

Traps may be used for verticals in the same manner as we apply them in dipoles and beam antennas. A multiband trap vertical is, in my humble opinion, the worst performer of the group of popular reduced-size verticals. The greater the number of traps used, the higher the system losses.

The unfortunate fact of the matter is that a trap vertical has the high-band element at the bottom, then works progressively upward as the frequency is lowered. This means that the 10-meter section, for example, is at the bottom and the 15-meter section is next, and so on. In practice, we should have the 10-meter radiator the highest and the low-frequency section should be at the bottom. This poor arrangement is less than ideal, especially for ground-mounted trap verticals.

Because of the foregoing problem, we will fare better when we mount a trap vertical as high above ground as we can. This calls for above-ground radial wires directly below the driven element. You should have four radials for each band of antenna operation. This can become complicated, owing to the large number of wires for a multiband vertical. Some amateurs use only two radials for each band, but better performance can be expected with four radials per band.

Fig 4-17 shows how a trap vertical is structured. The coil and capacitance hat atop the vertical is usually referred to as a <u>resonator</u>. For example, a 40-through-10 meter trap vertical may be used on 80 meters by adding the top resonator. The overall system then becomes a top-loaded 80-meter vertical, with the various traps in the system acting also as loading coils for 80 meters. This arrangement provides very narrow bandwidth on 80 meters (on the order of 25 kHz). It is a compromise antenna at best. Most multiband trap verticals are only 25-30 feet high, and this degrades the antenna performance also -- especially at 40 and 80 meters.

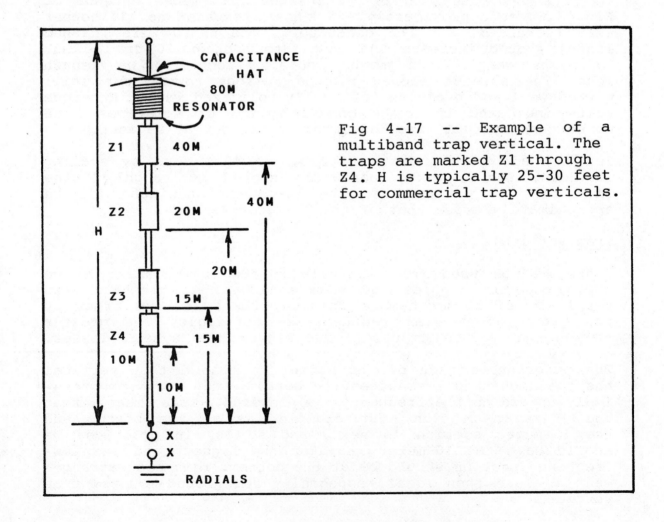

Fig 4-17 -- Example of a multiband trap vertical. The traps are marked Z1 through Z4. H is typically 25-30 feet for commercial trap verticals.

The working portions of the antenna in Fig 4-17 are specified in meters (M) to show that the high-band sections are toward the bottom of the antenna, as discussed earlier. This style of antenna is fed with 50-ohm coaxial cable. The traps are enclosed in aluminum tubes and rely on an inner aluminum tubing section to provide capacitance to the coil and outer tubing for ensuring trap resonance. In other words, there is no conventional fixed-value capacitor in parallel with the trap coil, as would be the case with a homemade coil-capacitor trap.

Chapter Summary

In this chapter I have described vertical antennas that are easy to build. You should make an effort to use a full-size vertical whenever possible. Don't skimp on the ground system, for the more extensive it is the greater your antenna efficiency. Your radials need not be deployed linearly below the antenna. You may have to route some of them around buildings and trees. The important consideration is that you use as many radials as you can manage. Even if some of them are substantially shorter than 1/4 wavelength, put them in or on the ground. They will help to improve the overall ground system. If you can install only a few radial wires, don't give up. You may be pleasantly surprised at the antenna performance even if the ground system is far from ideal. Tie everything that resembles a ground into your radial system, such as cold-water pipes, well casings and chain-link fences.

Chapter 5

HIGH-PERFORMANCE WIRE ANTENNAS

If you have sufficient real estate to erect antennas that are larger than dipoles and verticals, this chapter is for you. We will discuss a variety of antennas that are made from wire, and offer high performance compared to dipoles. These antennas are longer than 1/2 wavelength, and some of them provide gain over a dipole.

The rules concerning antenna performance versus height above ground apply to most of the antennas in this chapter. There-fore, you should make an effort to erect these antennas as high and in the clear as possible. However, if you can't provide the recommended antenna height for one or more of the radiators in this chapter, don't give up! Most of these antennas offer surprising performance, even at reduced height. If I may offer a simple truism, "You'll never know how well or poorly any antenna performs until you try it."

The Long-Wire Antenna

This antenna is not a true "long wire" unless it is one wavelength or greater in size. A long piece of wire (relative) may be a half-wavelength or random-length antenna, but it is not a long wire by definition.

You need to think of long wires in terms of one or more wavelengths. These antennas provide gain and many lobes of various radiation angles. Please refer to chapter 3 for a detailed discussion of the long wire. I mention it here to establish that most high-performance wire antennas are combinations of long wires or one-wavelength configurat-ions of the long wire.

Large Loop Antennas

You may erect a full-wave, closed-loop antenna parallel to ground, perpendicular to earth, or you may slope it from one or two supporting structures. A loop that is parallel to the ground will produce high-angle radiation. This is excellent for close-in communications, but the DX performance is anything but spectacular. The horizontal loop, as they are called, is generally less difficult to erect than is a vertical loop. This is because the vertical loop requires fairly high support poles or towers if it is to deliver optimum performance. Loops can be any convenient shape. Some illustrations of popular formats are provided in Fig 5-1.

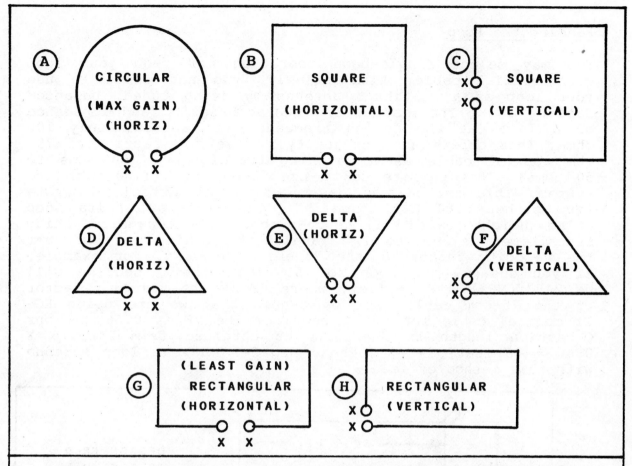

Fig 5-1 -- Various shapes for full-wave closed loops.
The circular format offers maximum gain over a dipole,
and the square loop comes next. Delta loops have less
gain than square ones, and the rectangular loop has
the least gain. The polarization is dependent upon
the point along the loop where the feed line is attached.
The polarization of these loops is indicated.

The overall conductor length for a full-wave loop is obtained
from L(feet) = 1005/f(MHz). Lower-corner feed (Fig 5-1 F
and H) results in vertical polarization with a low radiation
or launch angle, making these antennas ideal for DX work.
The effective height of a delta loop is best when the apex
is down (Fig 5-1E) as opposed to the arrangement at D of
Fig 5-1. Layout F requires only one tall support structure,
and the performance is better than the apex-up version at
D of Fig 5-1. If your tower or mast is not high enough to
support these loops perpendicular to earth, you may tilt
them away from the tower. Good results may be had when the
lower leg of a loop is only 6-8 feet above ground. But,
you should try to have the lower side at least 1/4 wavelength
above ground for best performance.

Full-wave loops may be used for multiband service if you
feed them with balanced transmission line (450-600 ohms).
These antennas will produce increasing gain as the order
of the harmonic becomes greater.

Feeding the Loop

You may select single-band operation and feed your loop
with coaxial cable. Alternatively, you can elect to use
your loop as a multiband antenna by using tuned, balanced
feeders and a Transmatch. The characteristic feed impedance
of a loop (at the design frequency) is approximately 100
ohms. This offers an opportunity to use a Q section of 75-
ohm coaxial cable as a matching transformer (100 ohms to
50 ohms). This permits us to use 50-ohm feed line with the
system. This method was described in an article I wrote
for October 1984 **QST**, page 24 ("The Full-Wave Delta Loop
at Low Height") with W1SE as co-author. The coaxial matching
transformer is cut to 1/4 wavelength, taking into account
the velocity factor of the 75-ohm line used. For example,
if you design your loop for 3.6 MHz, the Q section will
be 246/f(MHz) X VF = feet, where VF is the velocity factor
of the 75-ohm cable. Let us assume that we are using RG-
11 coaxial cable for the Q section. The VF is 0.66, so our
Q-section length in feet can be obtained from 246/3.6 X
0.66 = 45 feet, 1-1/4 inches. Fig 5-2 shows a loop antenna
with this method of feed.

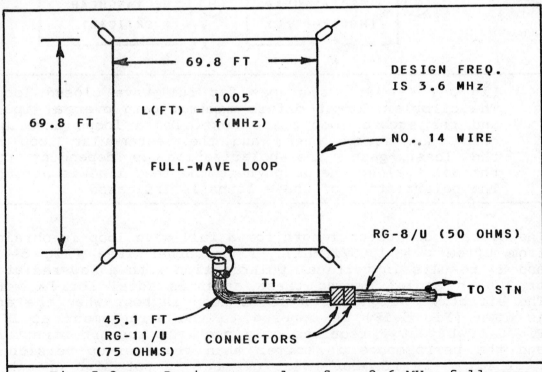

$$L(FT) = \frac{1005}{f(MHz)}$$

Fig 5-2 -- Design example of a 3.6-MHz full-wave
rectangular loop. T1 is a 1/4-wavelength matching
transformer to convert the 100-ohm feed point to
a 50-ohm coaxial feed line.

The 50-ohm feed line for your 80-meter loop may be any length that is convenient. You will find this and other full-wave loops to be broadband, compared to a half-wave dipole for the same frequency. A bonus feature of closed loops is their "quiet" nature during receive. They offer good immunity to man-made noise by virtue of their being closed circuits. Noise reduction of this type can be beneficial, especially during the reception of weak signals.

Loop Directivity

You will observe maximum bidirectional radiation from the loop of Fig 5-2 off the broad side of the antenna. Useful radiation will occur in the plane of the loop (off the sides), so don't worry about not being heard in the nulls of the loop. As the operating frequency is increased, the antenna becomes more directional off the sides, and a there can be as much as 5-6 dB of gain for high-order harmonic use. I had the interesting experience with a north-south 75-meter loop of finding it on par with, or up to 6 dB better, than my tribander on 15 and 20 meters. This was noted when working European stations from NW lower MI. The tribander was 55 feet above ground. My loop had the delta arrangement in Fig 5-1F, with lower-corner feed. Homemade 600-ohm open wire feeders and a 4:1 balun transformer were used to feed the loop. Fig 5-3 shows the arrangement I used. I attribute the exceptional DX performance to the low radiation angle of the loop with its lower-corner feed and vertical polarization. The theoretical launch angle of the antenna, respective to the horizon, is on the order of 10-15 degrees. The tribander at only 55 feet would have a substantially higher radiation angle.

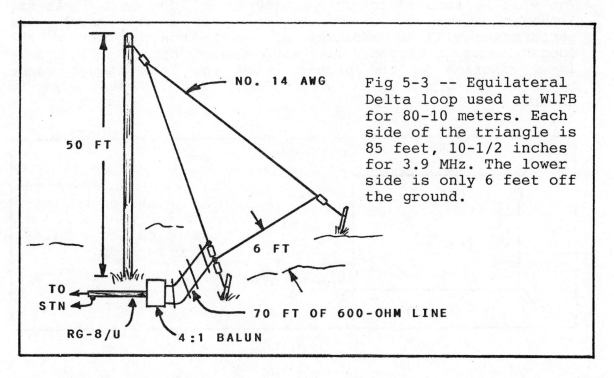

NO. 14 AWG

50 FT

6 FT

TO
STN

RG-8/U   4:1 BALUN   70 FT OF 600-OHM LINE

Fig 5-3 -- Equilateral Delta loop used at W1FB for 80-10 meters. Each side of the triangle is 85 feet, 10-1/2 inches for 3.9 MHz. The lower side is only 6 feet off the ground.

Although the low height of the lower side of the delta loop in Fig 5-3 is far from ideal, you may resort to this type of installation when you lack a support structure with ample height to elevate the bottom of the loop high above ground.

The balanced feed line in Fig 5-3 may be any length you choose. Also, you can use 300-ohm or 450-ohm ribbon line if you wish. The losses will be the lowest if you use open-wire feed line. The 4:1 balun transformer should be near your Transmatch. I have mine just outside the ham-shack window. The RG-8/U comes through the wall to my Transmatch.

If you are not interested in multiband operation with a delta loop, you may feed it in the manner shown in the drawing of Fig 5-2. You will notice that the delta loop in Fig 5-3 is tilted away from the mast. This does not appear to impair the performance, although a vertically erected loop should be your objective when possible.

If you are unable to find room for an equilateral triangle when erecting your loop, don't give up. One side may be considerably longer than the other two. This will reduce the loop gain slightly, but it will still give good performance. W8JUY erected a delta loop that had the apex down and the flat side up. The upper side (for 3.9 MHz) was more than 100 feet long, which resulted in two fairly short sides for the triangle. He fed his loop near ground, at the apex, as in Fig 5-1E. He reported excellent performance from 80 through 10 meters while using 450-ohm ladder-line feeders.

80-Meter Loops on 160 M

You will be tempted to try your 80-meter loop on 160 meters. Although it can be loaded by means of a Transmatch, the performance will be mediocre at best. This is because the loop becomes a closed, half-wave device at 1.8 MHz. There is a solution to the problem when you lack enough space to erect a full-size 160-meter loop. Fig 5-4 shows how to do this.

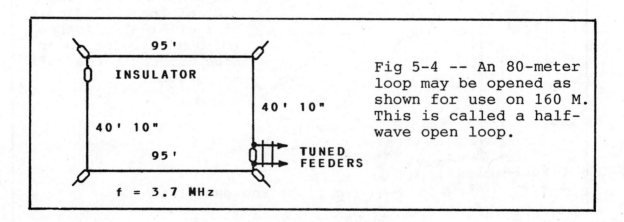

Fig 5-4 -- An 80-meter loop may be opened as shown for use on 160 M. This is called a half-wave open loop.

Although the antenna in Fig 5-4 is frequently called a half-wave open loop, it is not a loop, per se. It is a loop-shaped radiator, but a true loop is a closed circuit.

What we have in effect (Fig 5-4) is an inverted-L antenna for 160 meters, operating in combination with an L-shaped counterpoise. The left-hand vertical side of the antenna could be a straight extension of the lower side of the loop. It does not have to be bent upward as shown. But, this method of erection saves space, and therefore has merit. A relay could be located at the upper left loop insulator to open and close the loop for 160- and 80-meter operation. It could be controlled remotely from the ham shack. Please note that the loop is opened at a point that is <u>electrically</u> opposite the feed point.

I am presently using the antenna of Fig 5-4. The lower side of the loop is only 6 feet above ground. The system is strung between my 50-foot Rohn tower and a 50-foot telescoping mast. Not only does it work very well on 160 meters, it delivers excellent performance on 80, 40, 30, 20, 12 and 10 meters. I am using about 60 feet of 450-ohm molded feed line, a 4:1 balun transformer, then RG-8/U to the Transmatch, as in Fig 5-3.   Response to man-made noise (QRN) is somewhat greater with this antenna than when I used it as a closed loop. The polarization is vertical with the antenna of Fig 5-4. I do not observe any particular directionality with this system.

Half-Wave Delta Loop

I worked with Jack Belrose, VE2CV, in the practical development of a grounded Half Delta Loop. This work was described in **QST** for September, 1982, page 28. The antenna consists of a half wavelength conductor, of which part is made from wire and the remainder of the conductor is a metal tower or mast. The missing half of the full-wave loop is the image half of the antenna in the ground. Fig 5-5 shows how this antenna is configured.

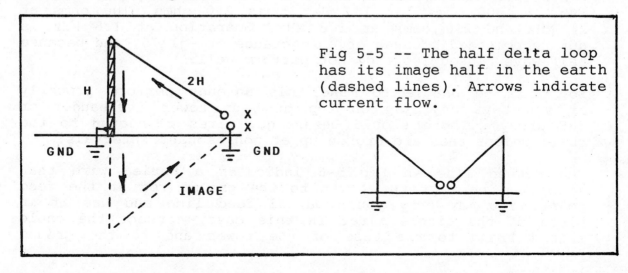

Fig 5-5 -- The half delta loop has its image half in the earth (dashed lines). Arrows indicate current flow.

The antenna in Fig 5-5 is vertically polarized and has an omnidirectional radiation pattern on its fundamental frequency. At harmonic frequencies the directivity is in the plane of the loop, but the polarization remains vertical. The gain over a dipole increases as the operating frequency is raised. The magnitude of gain may reach 5-6 dB at 28 MHz off the slope of the 2H wire (Fig 5-5).

Fig 5-6 illustrates a practical version of the half delta loop. It is resonant at 3.525 MHz with the dimensions given.

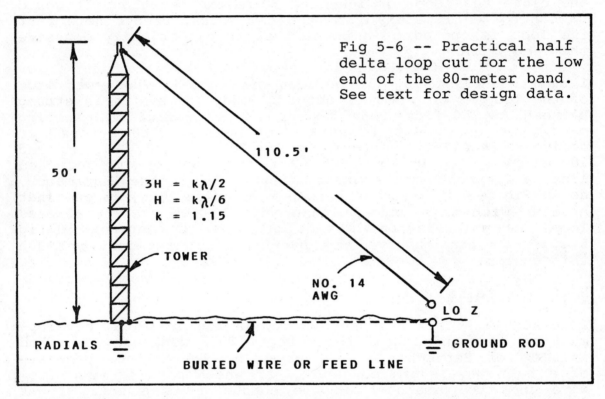

Fig 5-6 -- Practical half delta loop cut for the low end of the 80-meter band. See text for design data.

The feed impedance of the antenna in Fig 5-6 is on the order of 220 ohms at 3.525 MHz. At 4 MHz it drops to 140 ohms resistive. Measurements indicate that the impedance at 7 MHz is 1000 ohms, at 14 MHz it is 250 ohms, 100 ohms at 21 MHz and 290 ohms at 1.8 MHz. Operation at 1.8 MHz is not recommended because of a reactance of -j1175, and because half-wave closed loops do not perform well.

You will be unable to make this antenna perform properly if you have anything on top of your tower (tribander or VHF array). There should be no guy wires connected to the tower unless they are broken up at nonresonant intervals.

The dashed line in Fig 5-6 indicates a buried wire that joins the base of the tower to the ground rod at the feed point. You may bury your coaxial feed line and use it in place of the single wire. In this case, connect the cable shield braid to the base of the tower and to the ground rod. A radial system is required from the base of the tower in order to make this antenna function efficiently. Although a single ground post is indicated at the feed point, I found that performance improved when I attached four 1/4-wavelength radials to the ground post.

You will observe that the half delta loop is very quiet during receive, as is the situation with an above ground closed loop. This antenna has a signal launch angle of approximately 10° to the horizon, which makes it an excellent DX getter. An antenna of this type is easy to erect temporarily for Field Day use, and it permits multiband operation.

Two half delta loops may be combined as shown at the lower right in Fig 5-5. This results in additional gain, and the feed impedance will drop to approximately half that of a single half delta loop. You may want to consider building an antenna of this type if you have sufficient real estate to accommodate the double-delta configuration.

High-Angle Loop

Thus far we have dealt with loops that are perpendicular, or nearly so, to the earth. Earlier I mentioned the use of a full-wave loop that is parallel to earth. At the design frequency of this loop, the radiation is essentially straight up. This is ideal for close-in communications out to 500-600 miles at 1.8, 3.5 and 7 MHz. The benefits of high-angle radiation are realized at night, especially, when a reduction in QSB is desired. Fig 5-7 shows how you can arrange a horizontal loop.

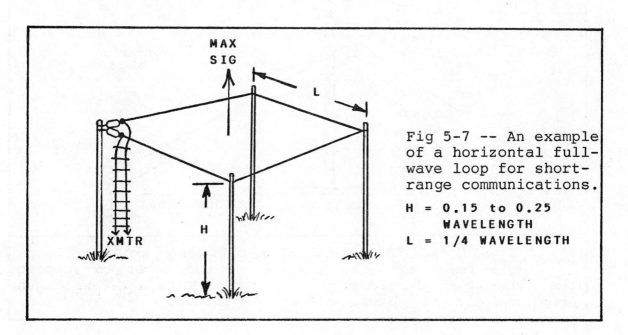

Fig 5-7 -- An example of a horizontal full-wave loop for short-range communications.

H = 0.15 to 0.25 WAVELENGTH

L = 1/4 WAVELENGTH

The loop in Fig 5-7 may be used on the harmonics of the fundamental frequency if you employ tuned feeders. As the operating frequency is increased you will find that the antenna begins to work better and better for DX communications, owing to minor lobes that have fairly low angles. But, at the fundamental frequency of the loop the maximum radiation is skyward, as shown by the arrow. Suggested height for the loop is 0.15 to 0.25 wavelength above ground. The legs of the square loop are 0.25 wavelength long, based on an overall conductor size obtained from L(feet) = 1005/f.

I had the occasion to use a very large loop of this type during my VP2MFW operation on Montserrat. It was 1000 feet long per leg, and was supported on 60-foot poles. A balun transformer (4:1) was attached to the feed point of the loop, and RG-8/U coaxial line ran from the balun to the operating position. This antenna exhibited a nearly perfect 1:1 SWR from 160 through 10 meters, and it worked very well for DXing on all of the HF bands. No Transmatch was necessary when QSYing from band to band.

Full-Wave Grounded Loop

If you have a tower that contains a triband Yagi and/or VHF antennas, you may use it to support a large grounded loop. The antennas on the tower will have minor effect on the system performance. Fig 5-8 illustrates an antenna of this kind.

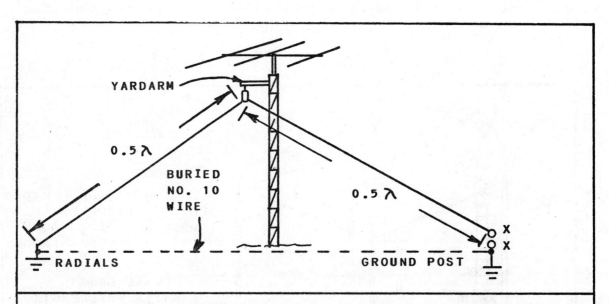

Fig 5-8 -- Method for erecting a full-wave grounded loop. A yardarm is attached to the top of the tower to support the midpoint of the loop. In effect, this is a 2-wavelength loop, owing to the image half in the ground.

I have had very good all-round results with the antenna
in Fig 5-8. As shown, it is a single-band radiator, since
balanced feed line is not suitable for feeding the system.
However, you may connect a 1:1 balun transformer to the
feed point, then use tuned feeders to the ham shack. This
requires a balun at each end of the tuned feeders. An alternative method you can apply for multiband use is to place
a switchable L network at the feed point for matching a
50-ohm coaxial feed line on various bands. Relays may be
used for selecting the coil taps, and a 1-RPM dc motor
can be included for adjustment of the tuning capacitor
in the L network. This will permit you to control the matching network from the operating position.

The higher the antenna midpoint the better the performance,
provided the enclosed angle is no less than 45 degrees.
This antenna has vertical polarization and it produces
lobes that have low radiation angles. As is the situation
with the half delta loop of Fig 5-5, there is an image
half to this antenna. Therefore, we may think of this loop
as a two-wavelength system, even though only one wavelength
of wire is used.

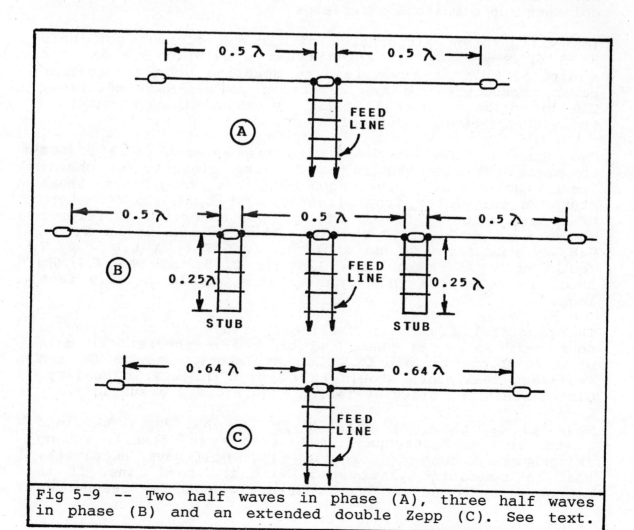

Fig 5-9 -- Two half waves in phase (A), three half waves
in phase (B) and an extended double Zepp (C). See text.

Collinear Arrays

We may combine half wavelengths of wire as shown in Fig 5-9 to obtain well-defined bidirectional radiation patterns. The combining of half-wave elements results in gain over a dipole type of antenna. The feed-point impedance of these antennas is high -- on the order of 5000 ohms when the elements are made from wire. VHF versions of these antennas, in which the conductor size (tubing or rods) is a greater fraction of a wavelength, have a feed impedance as low as 1000 ohms. In either case, we need to use low-loss open wire feed line of the 450-600 ohm class in order to minimize feeder losses.

The simplest and most common form of collinear array is that of Fig 5-9A. Here we have two half waves in phase. This antenna has a theoretical gain of 1.9 dB over that of a dipole. An 80-meter center-fed Zepp with tuned feeders, when used on 40 meters, is equivalent to the antenna of Fig 5-9A, by way of comparison. Maximum radiation is broadside to the antenna. Owing to the arrangement of the antenna elements, plus the tuned feed line, you may use a two-element collinear as a multiband radiator.

You will see an elaboration of the two-element collinear at B of Fig 5-9. This antenna has a gain of 3.2 dB over a dipole. This is equivalent to doubling your transmitter power, respective to the direction of maximum radiation. The broadside, figure-8 lobes are somewhat narrower than those of the two-element collinear.

Two phasing stubs are used with the antenna of Fig 5-9B to place the currents in all of the elements in phase. These stubs are 0.25 wavelength long. The length of these stubs is calculated from L(feet) = 246/F(MHz) x VF, where VF is the velocity factor of the stub material. Therefore, if we were to use 300-ohm TV ribbon, which has a VF of 0.8, a stub for 7.1 MHz would be (246/7.1) X 0.8 = 27.7 feet. The VF for open-wire line is 0.95, so the 7.1-MHz stub, if made from this type of line, would be 32.9 feet long.

The practical limit for collinear arrays of this type is four half waves in phase, which has a theoretical gain of 4.3 dB over a dipole. The insulators used in 3- and 4-element collinears should be of high dielectric quality. Glazed steatite, glass or Teflon insulators are suitable.

You can calculate the length of the half-wave sections of the collinear antenna from L(feet) = 468/f(MHz). Tuneup is generally done by erecting the half-wave section(s) that is immediately in contact with the feed line. If it is necessary to adjust the radiator lengths for resonance,

do this before adding the stubs and remaining collinear
sections. Performance will be good, even if the antenna
sections are not exactly resonant in your chosen part of
a band. The tuned feeders will permit you to maintain an
SWR of 1:1 via your Transmatch. Additional data on collinear
arrays is presented in **The ARRL Antenna Book.**

Although the "extended double Zepp" is not a true collinear
antenna, it performs in a similar manner. The details are
given in Fig 5-9C. This is a popular antenna among 80- and
40-meter operators. The gain over a dipole is on the order
of 3 dB, which effectively doubles the transmitter power
in the favored direction of the antenna radiation. Each
element is 0.64 wavelength long. A figure-8 radiation pattern
is characteristic of this antenna. The beamwidth is fairly
narrow compared to that of the two-element collinear in
Fig 5-9A. There are four smaller lobes that provide an approximate clover-leaf pattern. They appear at approximately
45° to the plane of the antenna. The six lobes allow good
coverage in several compass directions. You may calculate
the element lengths from wavelength = 936/f(MHz). Thus,
for an extended double Zepp cut for 3.55 MHz, each element
is 168.74 feet long, inclusive of the 0.64-wavelength factor.

Cloud-Warmer Antennas

During the discussion of Fig 5-7 we learned that a wire
antenna may be used to beam the signal skyward for effective
short-range communications in the MF and HF bands. I like
to refer to these antennas as "cloud warmers." Some hams
call them "scatter beams." They are highly useful for net
operation and ragchewing during after-dark hours for distances
out to, say, 600 miles.

You can build two-element wire beams for this purpose, and
they provide gain over a dipole. For example, a two-element
Yagi that has a driven element and a reflector (0.15-wave-
length element spacing) has a theoretical forward gain of
5.4 dB. A driven element and director, on the other hand,
can produce a gain of nearly 6 dB (0.12-wavelength spacing).
Therefore, if we construct a two-element wire beam for 75
meters, and aim it toward the sky, we end up with useful
high-angle radiation, plus gain.

Some amateurs configure their cloud warmers as two-element
inverted-V wire beams. That is, the upper inverted V is the
driven element. Another V-shaped wire (the reflector) is
erected 0.1 to 0.15 wavelength below the upper one. You
may install your two-element cloud warmer as two horizontal
half-wavelength wires, one above the other. The upper wire
is a dipole and the lower one is a reflector. This requires
two support poles, whereas the inverted-V format calls for
but one support structure. In either event, the reflector
is dimensioned approximately 5% longer than the element
that is driven (dipole). Fig 5-10 shows these antennas.

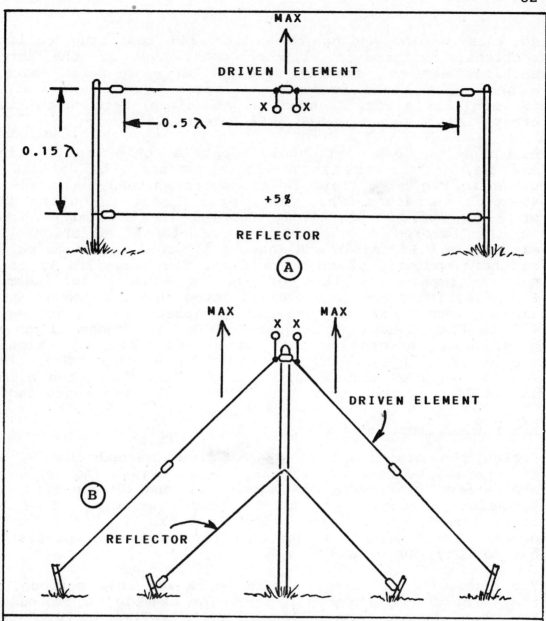

Fig 5-10 -- Examples of two-element "cloud-warmer" beams for high-angle radiation. Driven-element length is ontained from L(ft) = 468/f(MHz). The reflector is 5% longer than the driven element. These antennas provide in excess of 5 dB of gain.

The distance between the reflector and ground is not critical, but you should try to have both elements as high above ground as is practicable. Assume that you wish to design the antenna of Fig 5-10A for use at 3.8 MHz. The driven element will be 123.16 feet long. The reflector (5% longer) will be 129.3 feet long. If you use 0.15-wavelength spacing between the elements, the distance will be 38.8 feet. The feed impedance for antennas A and B of Fig 5-10 is on the order or 25 ohms. Therefore, you may wish to use one of the matching sections described later in this book in order to obtain a 50-ohm feed impedance. The hairpin match is my favorite for this type of antenna. It can be made from a section of open-wire or 450-ohm ladder line.

You may use the antennas of Fig 5-10 as conventional two-element Yagis by having both elements the same height above ground. In order for a wire beam antenna to be effective, it should be 0.5 wavelength above ground, although I have heard of some hams having good DX results when the antenna height was only 0.25 wavelength. The radiation angle would be much lower if the antenna were 0.5 wavelength high. A wire beam antenna is useful for DX work on 40 or 75 meters. It can be aimed at your favorite part of the world. The bandwidth of wire Yagis is quite narrow compared to the bandwidth of typical triband Yagis. This is because of the small diameter of the antenna elements. It is for this reason that the dimensions should be optimized for the part of the band where your DX activity will take place. You may increase the antenna bandwidth by using two or more fanned wires for each half of each element.

Chapter Summary

I have described antennas that I consider top performers versus cost and complexity. I have had good results with each antenna covered in this chapter. I have no reservations about recommending them to you.

There are numerous other high-performance wire antennas, such as the V beam and rhombic types. They are, however, somewhat beyond the scope of this text. Please refer to **The ARRL Antenna Book** for coverage of these and other high-performance wire antennas.

Chapter 6

LIMITED-SPACE and "INVISIBLE" ANTENNAS

There are occasions when we may not wish to erect an antenna that is easily observed by our neighbors. Perhaps we may adopt such a philosophy for aesthetic reasons. Whatever our objectives may be, there are a number of "invisible" antennas that we may use with effectiveness.

Perhaps you are an apartment dweller or the owner of a very small city lot. A situation of this type can often dictate a need for small antennas for Amateur Radio communications. Again, there are physically short antennas that we may use with good results. I want to mention, however, that most of these compromise antennas do not offer the performance of full-size equivalent antennas. Generally speaking, the antenna gain will be less than that of a full-size dipole, and the bandwidth will be narrower than that of a larger antenna. But, it is better to have a compromise antenna than no antenna at all. Furthermore, locating a suitable station ground may be difficult, especially if you live on an upper level of a condominium or apartment building. We will examine many of these problems in chapter 6, and solutions will be provided whenever possible.

Invisible Antennas

On several occasions I found it necessary to use an antenna that would not be easily detected by the casual onlooker. I found a quick solution to my problem through the use of light-gauge magnet wire. I have used 120-foot (and shorter) lengths of wire between a bedroom window and a nearby tree or other support structure. No. 26 AWG enameled wire has worked nicely for me, and it stays aloft if there is no icing or particularly strong wind. A bird may fly into your invisible antenna and break it, but these are the misfortunes of those of us who must employ an invisible wire antenna. Fig 6-1 shows this and other antennas that may be used when we do not wish to notify the world that we are Amateur Radio operators.

Invisible wire antennas can be made more obscure if we paint them with gray, light blue and black paint. Alternate the colors every six or eight inches. Spray-can paint may be used, or you can apply the paint with an artist's brush from paint cans. Exterior enamel paint is best for greater longevity. The end insulators can be made from monofilament fishing line, 8- or 10-pound test rating.

85

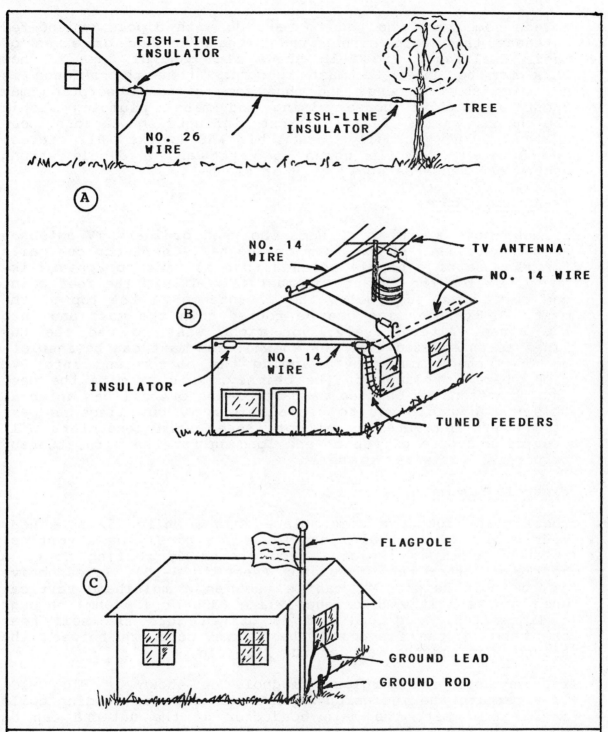

Fig 6-1 -- Examples of "invisible" antennas. A hard-to-see no. 26 wire is used at A as an end-fed antenna. Monofilament fish line loops serve as end insulators. At B are two hidden antennas. One is a center-fed Zepp strung under the eaves, and the other is an end-fed wire that appears to be the guying system for the TV antenna. A flagpole may be used (C) as a vertical antenna. A base insulator may be used to isolate the pole from ground.

I know some hams who had to make do with dipole or end-fed antennas that were strung under the overhang or eaves of their dwellings. Fig 6-1B shows the general idea of how this may be done. We must recognize that the antenna is not high above ground, and that some signal absorption may occur because of house wiring and metal plumbing within the house walls. Despite these influencing factors, one amateur I knew actually earned his WAS (worked all states) on 80-meter CW while using 25 watts and a dipole strung around the house as shown at B of Fig 6-1.

TV-Antenna Scam

I have known a number of hams who used ordinary TV antennas as ham antennas, in one form or another. One technique calls for the use of a dipole antenna that is made to appear like guy lines for the TV-antenna mast (Fig 6-1B). The feed point can be located just below the TV antenna at the top of the mast. RG-8X feed line can be routed down the mast and into the house. Alternatively, you might want to use the guy lines as an end-fed wire (as shown). The mast can be insulated from the chimney by slipping the bottom end into PVC pipe where it clamps to the chimney. This permits the mast and TV antenna to be used as part of the overall ham antenna. This would enable you to use all four TV guy lines as part of the end-fed antenna. All of these extra conductors will provide a degree of top or end loading for the wire, thereby requiring less wire, overall.

Flagpole Scheme

What could look more commonplace than a white flagpole near your dwelling? A metal flagpole serves nicely as a vertical antenna if it is isolated from ground by resting it on a bottle or other base insulator (see Fig 6-1C). Irrespective of the pole height, it can be used as a multiband vertical when you feed it with balanced line (300 or 450 ohm) through a Transmatch. A ground rod is driven into the soil near the base of the flagpole, The feeder connects between the ground post and the base of the flagpole.

You may use a fiberglass flagpole as shown in Fig 6-1C. This permits the inclusion of antenna traps or loading coils inside the mast. The main conductor of the antenna can be made from no. 10 copper wire or 1/4-inch copper tubing. Buried radials will aid the antenna performance greatly.

If you do not wish to "operate" on your fiberglass flagpole, as just suggested, you can use aircraft-grade steel cable as the halyard for your flag. This halyard then becomes the main conductor for the vertical antenna.

Other Obscure Antennas

Metal gutters and downspouts are sometimes used as amateur antennas. They constitute a random-length end-fed radiator. It is necessary to ensure that all joints between gutter and downspout sections have good electrical joints. You may do this by placing jumper straps across the joints. The straps are affixed to the gutter sections by means of no. 6 sheet-metal screws. Be sure to replace the lower parts of the downspouts with modern plastic downspout stock. This will prevent people and animals from coming in contact with the antenna when you are transmitting. You may connect your feed line to the lower end of one of the downspouts. Here again, you can use balanced, low-loss feeders. One side of the line connects to the downspout, and the remaining feeder wire is connected to a ground rod or ground system.

Clotheslines have served as invisible antennas for many hams. I once used an M-shaped antenna that was strung between four back-yard clothespoles. Each leg of the M was 16 feet long. I fed the antenna at one end of the M (bottom of one leg) through a Transmatch. The antenna was only 7 feet above ground, but it worked very well on 20, 15 and 10 meters. In used no. 10 vinyl-covered house wiring as the "clothes-line." Not only did it serve well as my antenna, but my XYL found it satisfactory as a wash line! Insulators were used at each clothes pole tie point. I threw a small cloth rag over the insulators to hide them, but nylon cord could have been used as insulator material. This would eliminate a need to hide the insulators. I worked a lot of DX with my clothes-line antenna over a two-year period.

Half-Sloper Antenna

An effective limited-space radiator is the "half sloper" or quarter-wave sloper. It requires considerably less real estate than does a horizontal quarter-wave antenna. The radiator is fed at the top by means of 50-ohm coaxial cable. Fig 6-2 shows a single-band version of this antenna. A metal mast or tower is used to support the antenna. The conduct-ive support member is a working part of the antenna because the shield braid of the feed line is connected to it near the antenna feed point. The bottom of the mast or tower is grounded. In effect, the system works somewhat like an inverted V (see chapter 2), but with a very small enclosed angle (typically 45°).

The half sloper provides vertical polarization. Maximum directivity is off the slope of the wire. The pitch angle and wire length is varied to obtain an SWR of 1:1. The mast or tower should not have guy wires that are resonant near the half-sloper resonant frequency. Break up your guy wires into nonresonant lengths by inserting strain insulators at appropriate intervals.

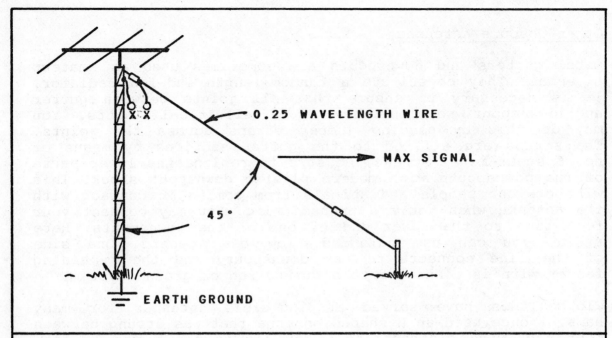

Fig 6-2 -- Example of a single-band half sloper. The antenna is fed at points X with 50-ohm coaxial cable. The radiator is 0.25 wavelength long. Tape feeder to tower, then vary enclosed angle and antenna length to obtain an SWR of 1:1.

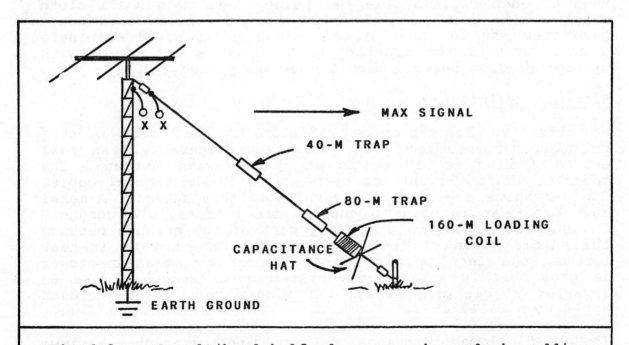

Fig 6-3 -- A multiband half sloper can be made by adding traps and an end-loading coil. This version operates on 160, 80 and 40 meters. Additional traps and antenna sections may be added for coverage of more HF bands.

If you do not have a metal support structure, you may erect your half sloper by attaching it to a tree or other nonconductive object. It will be necessary, however, to use a drop wire from the feed point to ground. This wire will take the place of the metal tower or mast in Fig 6-2. The bottom end of the drop wire must be grounded, and the shield braid of the coaxial feeder needs to be attached to the upper end of this wire.

Multiband Half Sloper

You will note in Fig 6-3 a three-band version of a half sloper. Two traps have been added. One is for 40 meters and the other is resonant in the 80-meter band. A 160-meter loading coil and capacitance hat has been added at the lower end of the slope wire.

It may not be possible to obtain an SWR of 1:1 on all three bands, but the SWR will be low. I used this antenna for two years and had good results with it. The SWR on 40 meters was 1:1. On 80 meters it was 1.4:1 and on 160 meters it was 1.8:1. I varied the angle and the length of the first wire section for a 1:1 SWR on 40 meters. The tower height was 50 feet. I found it necessary to run part of the 80-meter wire section and the 160-meter loading coil parallel to ground in order to get the necessary enclosed angle for a low SWR on 40 meters. The portion of the antenna that was parallel to ground was approximately 5 feet above ground.

Techniques for Antenna Shortening

When I refer to "antenna shortening" I am thinking about physical shortening. The antennas under discussion here are the correct electrical length, such as 0.25 and 0.5 wavelength.

Perhaps the most obvious method for decreasing the length or height of an antenna is to place a loading coil somewhere along the span of wire or other conductor. The helically wound antennas described earlier in this volume are good examples of shortened, resonant antennas.

Fig 6-4 shows some practical examples of shortened antennas. The dipole at A uses center-loading coils to achieve resonance in the desired part of the band. This antenna suffers from narrow bandwidth and reduced efficiency. It is, however, a practical antenna for small space, and it is worth considering as an indoor attic antenna.

A better scheme is illustrated in Fig 6-4 at B. Here we have added some stubs that are made from 300- or 450-ohm molded line. The length of these stubs should be chosen so as to not be resonant in any of the operating bands of the antenna. It is better to use several short stubs than a small number of longer ones. The stubs become part of the radiator, but yield an antenna that has its main part shorter than a full-fledged dipole. The bandwidth of this antenna is greater than that at A of Fig 6-4.

Other shortening methods may be used, such as those seen at C and D of Fig 6-4. You should try to be innovative when you design a shortened antenna. The configuration you select will need to be founded on the shape and size of the area you have available.

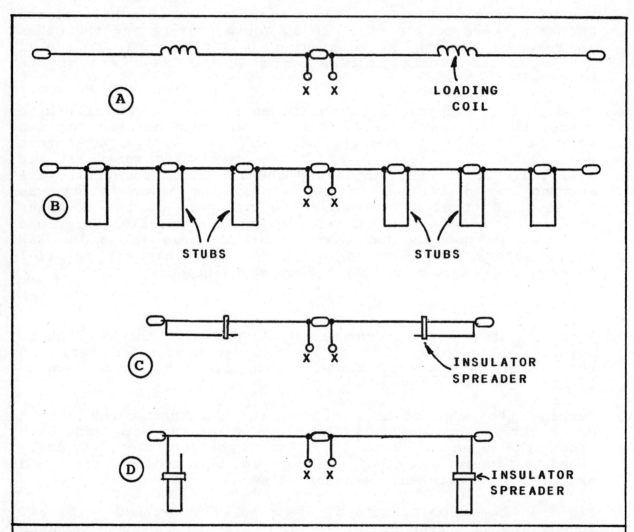

Fig 6-4 -- Examples of physically shortened dipoles. These techniques may be applied to verticals and end-fed antennas also. Loading inductors are used at A to reduce the antenna length. A group of stubs is used at B to load the antenna. They may be fashioned from 300-ohm TV ribbon, 450-ohm ladder line or open-wire line. Antenna C has the ends folded back, and the example at D has bent and folded ends. Each antenna is an electrical half wavelength.

Indoor Antennas

Although some of the antennas depicted in Fig 6-4 are suitable
for indoor use, they may be too long for the space you have
available. This can be especially germane to the matter if
you live in an apartment or small home.

VHF and UHF operation seldom causes a problem when we use
indoor antennas, but HF-band operation -- especially at the
low end of the spectrum -- can become a tough nut to crack!
I have developed a few small antennas for indoor use, and
I like to refer to them as "widgets." You will experience
fewer problems if you work with a short antenna that is an
electrical half wavelength. This type of antenna does not
require a ground as part of the working system. On the other
hand, a 0.25-wavelength radiator depends upon a good ground
system for proper performance. Obtaining a suitable ground
on a second or third floor of a dwelling is no simple task.
At 10 and 15 meters we can place a couple of 0.25-wavelength
radial wires on the floor, and this often solves our problem.
But, at 40 and 80 meters, for example, this approach may
be impossible. Therefore, it is better to use a 0.5-wavelength
radiator, however short it may be.

RF Energy in the Shack

The greatest problem you will experience when using indoor
antennas near your rig is stray RF energy. Modern transceivers
are sensitive to RF currents that enter the circuit. This
is due in part to the use of solid-state devices that work
all too well as rectifiers, even though they are supposed
to be amplifiers. The most common malady is RF energy in
the speech-amplifier stages of the transmitter. The mic cord
acts as a pickup antenna and routes unwanted RF energy into
the speech stages. This causes squeals and distortion. Some
brands of ham gear are worse than others in this respect.
If we use 0.25-wavelength antennas with an inferior ground
system, we are likely to be plagued by this problem. In a
like manner, end-fed 0.5-wavelength antennas bring high levels
of RF voltage near the rig. In any event, try to keep the
radiating portion of your indoor antenna as far from your
operating position as practicable. Connect your station gear
to the cold-water pipes and heating ducts if possible. This
will increase the effectiveness of the station ground. Stray
RF energy can also disrupt the operation of electronic keyers,
VFOs and other station equipment. RF energy on the mic lead
can quickly turn a speech amplifier into a "screech" amplifier!

Ideas for Widgets

Fig 6-5 shows some suggested designs for reduced-size indoor
antennas. I have used similar arrangements a number of   times
with acceptable results.

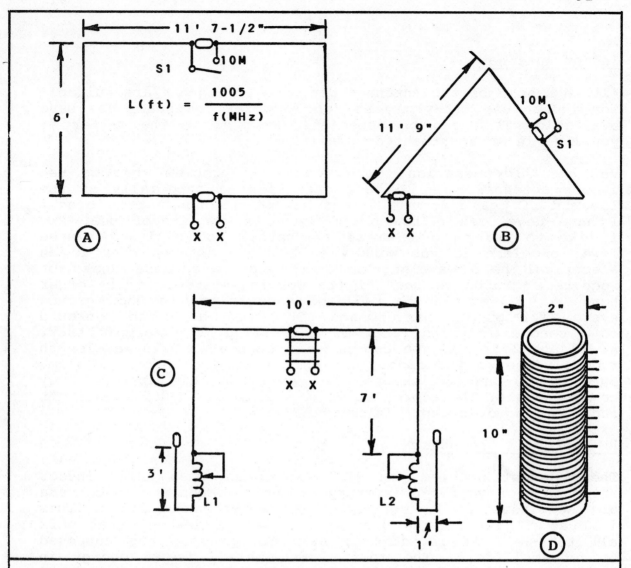

Fig 6-5 -- Indoor antennas that work well for their size. Antenna A is a full-wave loop for 28.5 MHz when S1 is closed. The antenna becomes a half-wave open loop on 14 MHz when S1 is open. Tuned feeders and a Transmatch are used with this loop. It will also perform on 15 meters, and may be used with some success on 40 meters when routed through a Transmatch. An equilateral triangle type of loop is shown at B. It may be used in place of antenna A. The dipole at C is suitable for use from 3.5 through 30 MHz if tuned feeders are used with a Transmatch. L1 and L2 are adjustable end-loading coils (D). Taps are used for changing bands. The dipole is resonant on 20 meters with the coils shorted out. A short lead with an alligator clip can be used at each coil for tap selection. The correct taps are selected by checking antenna resonance with a dip meter. The coil has 120 turns of no. 14 enamel wire, close wound on a 2-inch diameter X 10-inch low-loss coil form. The taps for various bands and segments of bands may be marked with a dab of paint (different colors) for quick identification when changing bands.

Chapter Summary

This chapter on limited-space antennas is by no means all inclusive. I have presented some antennas that I have used with success over the past few decades. Some of these antennas perform better than others described herein, but all of them permit solid QSOs when band conditions are average or better. You will be able to work DX with these antennas, part of the time.

I cannot stress strongly enough the principle of **Safety First**. Do not install any antenna so that humans and animals can come in contact with it. Indoor antennas should be located away from the main area of movement within the room where they are placed. The antennas of Fig 6-5, for example, can be affixed to the wall of the room with tape or small staples. Try to keep them as far from electrical wiring as possible. This will minimize TVI and RFI that is transmitted via the power wiring, and it will reduce losses caused by RF absorption by the electrical wiring.

A Transmatch will enable you to load almost anything that is conductive. Metal beds and bed springs, for example, have been used by many hams as indoor antennas. Window screens that are insulated from the steel framework of buildings can be used as antennas. I once knew a ham in Detroit who used his metal workbench as an antenna. He also unplugged his floor-model drill press from the wall outlet and used it as an indoor 10-meter antenna. Another ham that I knew lived in an apartment in NY City. He used the metal fire escape outside his window as a 40-meter antenna. He worked a substantial number of DX stations with that antenna!

Your imagination is the limiting factor when trying to survive as an apartment-dweller ham. There are many things that may be used as antennas, consistent with safety precautions. The important consideration is that you do not give up in despair when you aren't able to erect an outdoor antenna. Remember that "a poor antenna is better than no antenna!"

Chapter 7

MATCHING TECHNIQUES

Maximum power transfer cannot occur unless the impedances
between the mating circuits are matched. This refers to the
match between an antenna and the feed line, between the transmitter and the feeder or between the exciter and the power
amplifier. In an ideal situation the SWR between circuits
that are joined is 1:1. This indicates a matched condition.
The greater the mismatch, the higher the power loss in the
system.

A notable departure from this rule is when we feed a center-
fed Zepp dipole with balanced feeders. The feeders may have
an impedance between 300 and 600 ohms, depending upon the
conductor diameter and spacing, but the feed impedance of
the antenna may be somewhere between 50 and several thousand
ohms, depending upon the chosen operating frequency. Needless
to say, the transmission line may radiate because of high
standing waves along it. The overall system loss is minimal,
owing to the low-loss characteristic of the balanced feed
line. This is especially true when we use open-wire feeders,
which have the least loss of the common lines. If, on the
other hand, we fed the same antenna with coaxial cable, the
losses in the feed line would be quite high, especially at
the upper end of the HF spectrum. The longer the feed line
the greater the loss.

If we are to obtain maximum performance from any antenna
we use, we should make an effort to pay attention to proper
matching. This means we need to match each end of the feed
line to its load. It is not enough to create a match between
the transmitter and feeder (via a Transmatch), because the
antenna and feed line may still be mismatched. Matching at
the transmitter does not correct for a mismatch at the antenna.
The transmitter is, however, able to deliver its maximum
power into the feed line when a Transmatch is used in the
shack.

Common Matching Networks

Fig 7-1 shows some popular matching schemes for HF, VHF and
UHF antennas. They may be applied to a single radiator,
or they can be used at the driven element of a beam antenna.

A delta match is illustrated at A of Fig 7-1. The driven
element is a continuous conductor. The delta match is fanned
at the center of the radiator, then tapered to a feed point.
Dimensions "a" and "b" are changed until a match to the

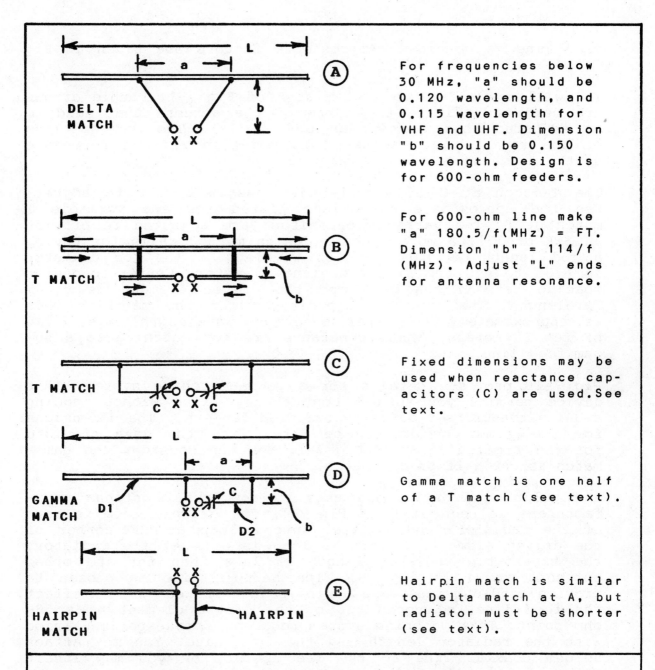

For frequencies below 30 MHz, "a" should be 0.120 wavelength, and 0.115 wavelength for VHF and UHF. Dimension "b" should be 0.150 wavelength. Design is for 600-ohm feeders.

For 600-ohm line make "a" 180.5/f(MHz) = FT. Dimension "b" = 114/f (MHz). Adjust "L" ends for antenna resonance.

Fixed dimensions may be used when reactance capacitors (C) are used. See text.

Gamma match is one half of a T match (see text).

Hairpin match is similar to Delta match at A, but radiator must be shorter (see text).

Fig 7-1 -- Examples of common matching systems for antenna elements made from aluminum tubing, such as those used in HF and VHF Yagi beam antennas. A delta match is shown at A. It is suitable for use with 300 to 600 ohm balanced feeders. A T match is illustrated at B and C. The gamma match at D is essentially one half of the T match shown at C. A hairpin matching device is found at E. The end-to-end radiator length is less than for the other antennas in this figure, owing to the hairpin being part of the radiator.

feed line is obtained. Approximate factors are given in Fig 7-1A.

You will see a T-match system at B of Fig 7-1. Double arrows are used to indicate the parts of the antenna that need to be adjusted when seeking an SWR of 1:1. The end sections of the radiator are adjusted to maintain antenna resonance as the matching section is adjusted.

The T-match at C of Fig 7-1 is less difficult to adjust. Use the approximate dimensions listed for the variable T match at B. The variable capacitors are adjusted to provide an SWR of 1:1. A more precise design method calls for having a rod length of 0.04 to 0.05 wavelength. The rod diameter is 0.33 to 0.5 the driven-element diameter. Center-to-center spacing between the rods and the driven element is 0.007 wavelength. The required capacitance for the matching rods is approximately 7 pF per meter of wavelength, i.e., 140 pF for 20 meters. These constants are for matching to a 50-ohm feed line.

Fig 7-1D provides details for a gamma-matching system. This device permits you to use coaxial feed line without adding a balun transformer or using balanced feeders. The dimensions for the gamma system may be taken from the data provided for the T match at C of Fig 7-1. You may consider the gamma match as one half of a T match.

A popular matching device used by commercial antenna manu-facturers is seen at E of Fig 7-1. The hairpin loop is part of the radiator, and is bent into a loop at the center of the driven element. Since the loop is part of the radiator, the tip-to-tip radiator length is less than for the other antennas shown in Fig 7-1. The hairpin size is chosen to provide a 50-ohm impedance at the points marked "X." In effect, the feed line is tapped out on the driven element at a 50-ohm point. This matching system requires some experimentation with the radiator length and the loop size. You may attach 50-ohm coaxial line to the feed point, or you may insert a 1:1 balun transformer between the coaxial line and the radiator feed point.

Methods for Stub Matching

A section of balanced transmission line can be used as a matching device when it is connected to the antenna feed point. The feed line is tapped on the stub at a point which provides a matched condition. The stub length may be 0.25 or 0.5 wavelength, depending on the antenna feed impedance. Also, the stub may be open or shorted at the bottom end. Some examples of matching stubs are shown in Fig 7-2. This type of matching device has been referred to by some amateurs as a "universal matching stub." It is easy to adjust and is inexpensive.

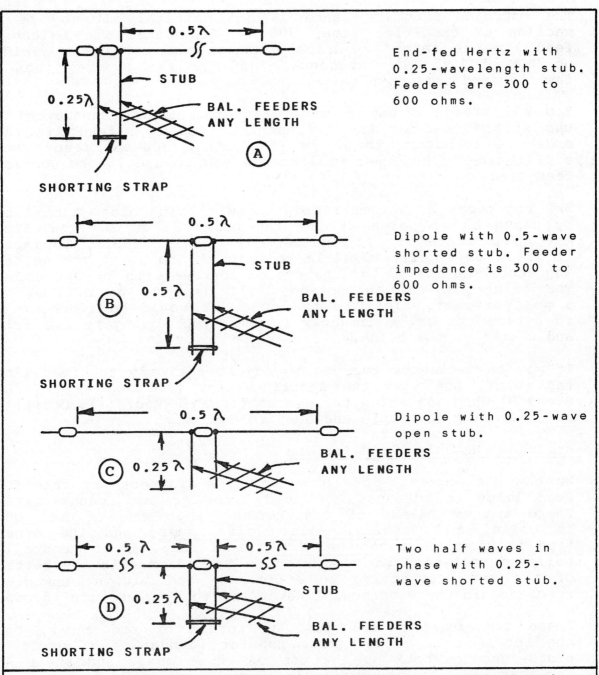

Fig 7-2 -- Various methods for applying a matching stub when using balanced, nonresonant feed line. The Feeders are moved along the stub until an impedance match to the transmission line is obtained. A 4:1 balun transformer may be used with 300-ohm feed line to permit feeding the system with 75-ohm coaxial cable. Alternatively, the 4:1 balun can be connected directly to the stub and moved along it to obtain a match to 50-ohm coaxial line. At A, B and D it may be necessary to move the shorting strap while moving the tap point for the feed line.

The matching stub impedance is not critical. It may be a section of open-wire line, 300 to 600 ohms. The balanced feed line is tapped at a position on the stub that yields an SWR of 1:1. The impedance along the stub varies from a few thousand ohms down to zero.

You will need to use a balun transformer or Transmatch at the station end of the nonresonant, balanced feed line in order to transform to 50 or 75 ohms. A 300-ohm feeder and a 4:1 balun transformer will enable you to use 75-ohm coaxial feed line.

SWR tests may be accomplished (while tapping along the stub) by using a 4:1 balun at the tap point. Your SWR indicator may be connected to the balun with coaxial cable. In fact, if it is physically possible to keep the balun at the matching stub, you may adjust the system for use with 50-ohm cable. The balun, in this situation, will need to be affixed to a pole or mast, close to the stub. It would be impractical to attach the balun directly to the stub, owing to the bulk and weight of the balun.

It may be necessary for you to adjust not only the feed-line tap point, but also the shorting-strap position (Fig 7-2A, B and D) when adjusting the system for an SWR of 1:1. Overall, stub matching is simple and easy to deal with.

Broadband Matching Transformers

We can use magnetic-core broadband transformers for matching feed lines to antennas, or feed lines to our transmitters. There are two kinds of transformers that we may use. One is called a transmission-line transformer, and the other type is known as a conventional transformer. It is generally believed that the transmission-line transformer is the better of the two, respective to efficiency and minimum unwanted reactance in the windings. I have used both types with success.

These transformers are wound on toroid or rod cores, but the toroid core is the most popular foundation. I wrote a plain-language book on the use of these cores and their RF applications. It is available from Prentice-Hall, Inc. or Amidon Assoc., Inc. The book title is **Ferromagmetic Core Design & Applications Handbook**, no. 0-13-314088-1. You will find this a useful reference if you plan to work with toroids, rods, sleeves and pot cores. Another book on this general subject is available from The ARRL, Inc. It was written by Jerry Sevick, W2FMI.[1]

Fig 7-3 shows a number of transformers that can be used for impedance matching in antenna systems. Some are for use in unbalanced-line applications. Others may be used between balanced and unbalanced terminations.

[1] **Transmission Line Transformers, W2FMI**

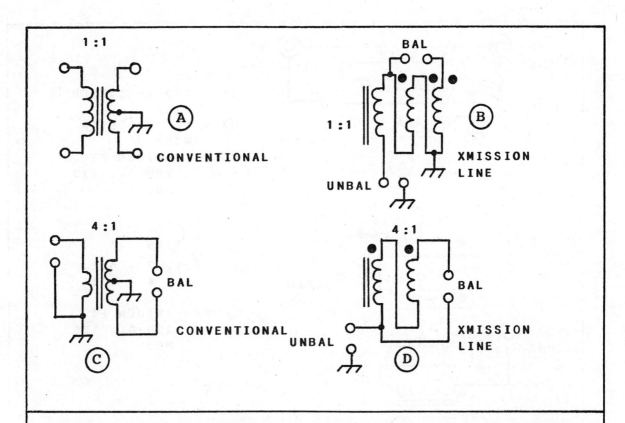

Fig 7-3 -- Illustrations of conventional and transmission-line broadband transformers that may be used in antenna systems. The black dots above the windings at B and D indicate the winding polarity or phasing. If correct phasing is not used, the transformers will not function. These transformers are wound on large ferrite or powdered-iron toroid or rod cores. The core size and wire diameter must be large enough to accommodate the transmitter power without excessive heating (see **The ARRL Handbook** or the W1FB book referenced earlier). These transformer windings require an inductive reactance of approximately four times the lowest terminal impedance. Thus for a 50-ohm load, the XL of the winding must be at least 200 ohms at the lowest proposed operating frequency.

Ferrite cores are best for use in broadband RF transformers. They have higher permeability for a given core size than is the case with powdered-iron cores. This permits you to use fewer turns of wire (desirable) for a given winding XL. This reduces losses and unwanted capacitive reactance. Ferrite cores will, however, saturate more readily than will powdered iron of equivalent cross-sectional area. Thus, for a specified RF power level, the ferrite core is generally larger than the powdered-iron core for the same type of transformer. For a discussion of balun-transformer design and practical applications see "How to Build and Use Balun Transformers," **QST,** March 1987, page 34.

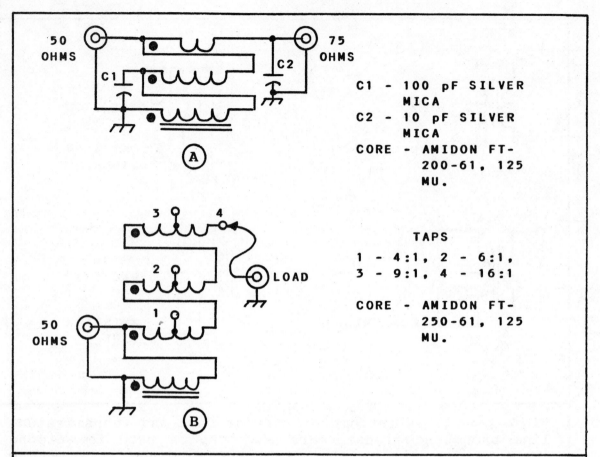

C1 - 100 pF SILVER MICA
C2 - 10 pF SILVER MICA
CORE - AMIDON FT-200-61, 125 MU.

TAPS
1 - 4:1, 2 - 6:1,
3 - 9:1, 4 - 16:1

CORE - AMIDON FT-250-61, 125 MU.

Fig 7-4 -- A broadband transformer that transforms 50 ohms to 75 ohms is shown at A. Use 6 trifilar turns of no. 14 enameled wire. The small winding has 3 turns. The black dots indicate polarity or phasing. The transformer at B was designed by W2FMI. It offers four transformation ratios. It may be used to step up or step down by reversing the input and output connections. As a stepdown transformer it can look into loads as low as 3-4 ohms. Use 10 quadrifilar turns of no. 14 enameled wire. These transformers are suitable for use from 1.8 to 30 MHz. Transformer B is useful for matching to low-impedance vertical antennas.

The transformer of Fig 7-4A is handy for interfacing 50-ohm coaxial cable to 75-ohm transmission line of the CATV Hardline variety. It was designed by W1VD, a former ARRL staff member.

The W2FMI variable-impedance broadband transformer of Fig 7-4B is especially useful for antenna experimentation and design. The leads for this transformer must be kept as short as practicable in order to minimize stray reactances, which can degrade the transformer performance. Both units in Fig 7-4 are rated at 1.5 kW PEP.

Transmatches

The question often arises about whether or not we need to use a Transmatch. You may know this device by another name, such as "antenna tuner" or "antenna matcher." It is also called an ATU (antenna tuning unit). The term Transmatch was coined by the late George Grammer, W1DF, who served for many years as the ARRL Technical Director. It is an apt name for the network, since it matches the transmitter to the feed line. Were it located at the antenna feed point, it would then be an antenna matcher.

Transmatches consist of variable elements of capacitance and inductance. These components are used to cancel inductive or capacitive reactance that may be present at the end of a feed line or at the input to an end-fed antenna. Our Transmatch enables us to provide a resistive (nonreactive) load for our station equipment, even though there may be an SWR as great as 8:1 on the feed line. This permits the transmitter to deliver its rated output power, and the receiver looks into a 50-ohm load to yield maximum signal response.

Most commercial Transmatches are single-ended units. That is, they are designed to be used in an unbalanced feed line, such as RG-8. In order to use a Transmatch with balanced feeders, such as 300-ohm ribbon, we must insert a 1:1 or 4:1 balun transformer between the Transmatch output and the balanced line. Some Transmatches have the balun built in. You should be aware that lowering the SWR at the station end of the transmission line does not correct the mismatch at the antenna. Standing waves will remain on the feeder, and there will be losses in the line.

Most balun transformers will work well into loads as high as 500-600 ohms. At higher impedances they no longer function as broadband transformers, and are prone to excessive heating and arcing at high power. This is because the RF voltage is quite high above 600 ohms, and this causes core saturation. The high RF voltage will arc between the winding and the core in severe situations, thereby destroying the balun. The point I am making is that a balun is not a cure-all for matching problems.

A Transmatch is useful, for example, when we have an antenna with a narrow bandwidth. Suppose your 75-meter dipole has a bandwidth of 100 kHz between the 2:1 SWR points of the response curve. This is fine, as far as your transmitter is concerned. But, when you try to operate somewhere else in the band (say, on 80 meters), the SWR rises to 5:1. The Transmatch will permit you to use the antenna if you use it to tune out the reactance. This gives your transmitter a 50-ohm load, even though the SWR is still high on the line.

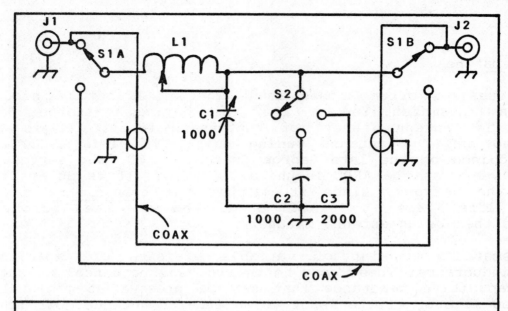

Fig 7-5 -- Schematic diagram of a reversible L-network Transmatch. Capacitance is in pF. C2 and C3 are transmitting ceramic or mica transmitting-grade capacitors. C1 is a variable capacitor with plate spacing appropriate for the power level of the transmitter. L1 is a rotary inductor or tapped coil (25 µH maximum inductance). J1 and J2 are SO-239 connectors or similar. S1 and S2 are ceramic rotary switches with contacts that can accommodate the power level anticipated.

L-Network Matcher

One of the simplest tunable matching networks we can build is the L type of circuit in Fig 7-5. There are only two adjustable elements, C1 and L1. The end of the L network that has the capacitors is the side that connects to the high-impedance load. For example, if we have a 50-ohm transmitter that must be matched to a 200-ohm antenna, we will have C1 on the 200-ohm side of the circuit. There are times, however, when we need to tranform from 50 ohms to some lower impedance, such as 50 ohms to 30 ohms. At such times we need to reverse the L network. This can be done by switching the connections to J1 and J2, but I have included S1 to make that job less difficult. S1 is the network reversing switch. Short runs of coaxial cable are used between the S1 switch sections, as shown in Fig 7-5.

C1, by itself, does not provide sufficient capacitance for matching low-impedance loads. S2 permits using 3000 pF of additional capacitance to increase the range of the matcher.

L1 of Fig 7-5 should have a maximum inductance of 20-25 µH in order to provide a wide matching range. A roller inductor provides the best resolution of inductance, but you may use a Miniductor or homemade coil to which you can add numerous taps. A third switch can be included for selecting the coil taps. In any event, all of the switches used need to be of high RF quality with ceramic insulation. The plate spacing for C1 must be ample for the RF power supplied to the matcher. The breakdown-voltage rating of C1 should be 1000 to 1500 volts for powers in excess of 500 watts.

You may include a 1:1 balun transformer with the circuit of Fig 7-5 if you plan to use the matcher with balanced feed line in the 300- to 600-ohm class. An SWR bridge is required between the transmitter and the matcher to provide a visual indication of the matcher tuning. Adjust the C and L values to secure an SWR of 1:1.

T-Network Transmatch

A flexible low-pass type of matcher that does not require a reversing switch is the popular T network. It contains two variable capacitors and one variable inductor. This tuner will accommodate a wide range of dissimilar impedances, from 50 ohms at one end to practically finite or infinite load impedances at the opposite end. In order to realize this wide matching range it is necessary to use a roller inductor: As little as one half a coil turn can mean the difference between a perfect or near-perfect match. Although you may use a tapped coil and a switch for varying the inductance of the matcher, you will encounter certain antenna impedances that a tapped coil will not resolve. The tapped-coil method is suitable if you work with the same group of antennas all of the time. In this situation you can experiment with the placement of the coil taps before making permanent solder connections on the coil. Changing frequency and switching antennas then becomes a routine task, respective to obtaining an SWR of 1:1.

A practical T-network circuit is presented in Fig 7-6A. This is a low-pass network, as is the L network of Fig 7-5. This means that the matcher will attenuate harmonics along with providing an impedance match. This can be helpful in the reduction of TVI and RFI. The amount of attenuation depends on the exact settings of the C and L constants, which are governed by the load presented by the antenna. The T network can, under certain load conditions, act as a high-pass network, which is not suitable for harmonic suppression.

A better circuit for the suppression of harmonics is shown at B of Fig 7-6. This is the SPC (series, parallel capacitance) Transmatch that I developed some years ago. The inductor has a lower parallel impedance than that of Fig 7-6A, owing to the shunt capacitance across the coil. This makes the tuning capacitors less prone to arc-over at high power, since the lower tank impedance results in lower peak RF voltage.

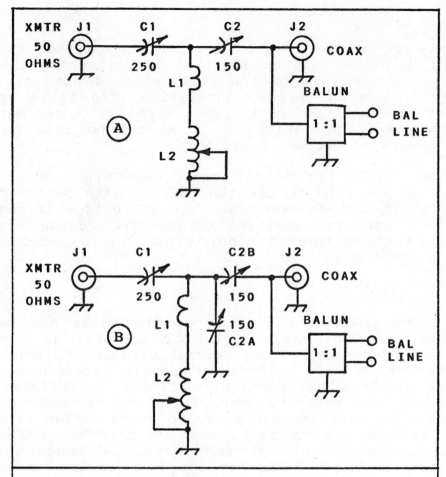

Fig 7-6 -- Practical circuits for (A) a T-network Transmatch and (B) an SPC Transmatch (modified T network). L1 is a 1-uH coil for use on 10 and 15 meters. Use 4 turns of no. 10 bus wire, 1 inch ID X 1-1/2 inches long. L2 is a 20 or 22-uH roller inductor. C1 may be a 200 pF capacitor. C2 at B is a split-stator variable with the rotor connected to the top of L1. Float the C2 frame above ground on standoff insulators. C1 and C2 of both circuits have their frames above ground.

The Transmatch in Fig 7-6A will operate from 3.5 to 30 MHz with the component values listed. Circuit B of Fig 7-6, because of the shunt capacitance of C2A, will permit you to use it from 1.8 to 30 MHz. L1 in both circuits is necessary because most roller coils do not have adequate turns taper to permit an inductance of high Q at 10 and 15 meters. Without L1 you might end up with only 1/2 or 1 turn of roller coil at those frequencies. You must also use very short leads in that part of the matcher circuit to minimize unwanted stray inductance

I suggest that you use 1/4-inch wide copper straps between the components (C1, C2, L1, L2, J1 and J2). This will help to reduce stray circuit inductance, and will therefore improve the matcher performance from 20 meters through 10 meters.

C1 and C2 of Fig 7-6 are mounted on insulating posts. Only the stator of C2A is grounded. Insulated shaft couplers are required between the tuning capacitors and the panel knobs. The roller coil does not need to be isolated from the panel and chassis.

Capacitor C2 of both circuits requires greater plate spacing than does C1. If flashover occurs, it will happen at C2. This will generally occur when the Transmatch is connected to a very low impedance, such as 20 to 75 ohms.

Adjustment of the three controls (Fig 7-6) is required most of the time when you change your operating frequency. You may start by setting each control at mid range. Then listen to the background noise or a weak signal in your receiver. Set the controls for maximum receiver response. Final tweaking of the controls may then be done with the transmitter delivering low power. Minimum insertion loss (wasted power) through the Transmatch will occur when you use the maximum amount of capacitance possible at C2, consistent with an SWR of 1:1. A variety of control settings can yield an SWR of 1:1, but try to select the settings that ensure high capacitance at C2.

Transmatch adjustment will be easier if you use vernier dials on C1 and C2. Under some matching conditions the tuning may be rather sharp. The vernier mechanisms can reduce this problem, and they provide you with logging scales for use when changing frequencies.

Chapter Summary

There are numerous circuits for matching networks. I have offered the ones that I feel are good performers for a wide range of impedances.

You will find other types of matching networks in **The ARRL Handbook** and also in the **The ARRL Antenna Book**. The theory of operation is explained more fully in those publications.

Chapter 8

SPECIAL RECEIVING ANTENNAS

We are often beset by man-made noise that can spoil reception, especially when we are attempting to copy weak stations. Our regular ham-station antennas may pick up too much local noise to permit us to hear   weak signals through the QRN. Vertical antennas are the worst type in this situation, and this is because most man-made noise is vertically polarized. There are a number of receiving-only antennas that we can build to minimize this problem. Small loops and very long Beverage antennas are used by amateurs for this purpose. Unfortunately, a Beverage antenna must be 1 wavelength or greater in size to perform in the classic low-noise, directional manner. A thorough discussion of this antenna, along with design data is provided in **QST**. The article was written by Beverage and DeMaw ("The Classic Beverage Antenna, Revisited"). It appeared in **QST** for January 1982, page 11.

Most of us lack the the space to erect a Beverage antenna. Alternatives exist for low-noise receiving antennas, and we shall discuss them in this chapter. These antennas lack gain, so it is necessary to use a low-noise RF preamplifier with them in order to ensure unity gain, or nearly so, with our transmitting antennas.

Small Receiving Loops

Small loops are usually 1/10th wavelength or less overall. They can be made from a single turn of coaxial cable, or from several turns of wire on an insulating frame. Some receiving loops are wound on ferrite rods or bars, which makes them very compact. The principle of operation remains the same, irrespective of the loop format.

The feed impedance of small loops is very low -- on the order of a few ohms. This makes them difficult to match to 50-ohm line, and losses are generally high. This is one reason why a preamplifier is needed. I recommend a preamplifier gain of 25 to 40 dB for all of the receiving antennas detailed here. A suitable circuit is offered in this chapter.

Some examples of receiving loops are given in Fig 8-1. I prefer the shielded loop of Fig 8-1B, since it helps to cancel some man-made noise by virtue of the outer shield.

Maximum signal response with these loops is in the plane of the loop, rather than off the broad side. Full-wave loops

operate in the opposite manner: Maximum radiation or response
is off the broad side of the loop. The exception for receiving
loops is when we use a ferrite-rod antenna. Maximum response
for this style of loop is off the broad side of the rod. In all
cases, the pattern is bidirectional. Deep nulls occur off
the nondirective sides of a loop, if it is well balanced or
symmetrical, electrically. Nulls as deep as 30-35 dB are common
with a well-designed loop. The loop bandwidth is quite narrow,
owing to the high Q of the antenna.

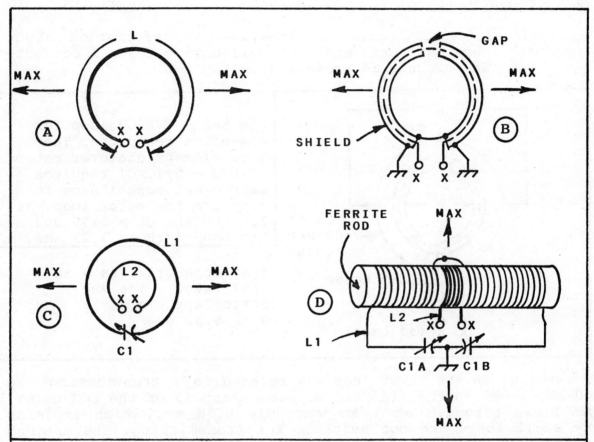

Fig 8-1 -- Examples of small loop antennas. Dimension
L at A is 1/10th wavelength. Maximum response is in
the plane of the loop, as shown by the arrows. Loop
B is electrostatically shielded. RG-8 cable or 50- or
75-ohm Hardline is excellent for this type of loop.
Coupling to the loop, and tuning, may be accomplished
as shown at C. The circuit at C shows how the loop can
be tuned to resonance (C1). L2 is a small coupling loop
within the main loop, L1. A ferrite-rod loop is shown
at D. L2 is the coupling link for the preamplifier.
An Amidon Assoc. 7-inch by 1/2 inch rod with a mu of
125 (no. 61 material) is suitable for use at 1.8 and
3.5 MHz. C1 tunes L1 to resonance. A series reactance
capacitor may be used with the coupling links at C and
D to provide matching to the receiver or preamplifier.
A high-capacitance trimmer (2000 pF maximum) is suitable
for 160 and 80 meters.

Receiving loops do not work especially well for DX weak-signal reception above approximately 2 MHz, although some amateurs have reported worthwhile results at 75 and 80 meters. A detailed article about loop design and use was published in **QST** for July 1977, page 30 ("Beat the Noise with a Scoop Loop"). Data is provided for building a shielded loop, plus a version that uses a ferrite rod. You may want to read that article. It offers design information about the loops in Fig 8-1.

VE2CV Concentric Loop

Jack Belrose, VE2CV, gave me some design information about a special loop that performed very well during tests I conducted in 1986. The circuit is presented in Fig 8-2.

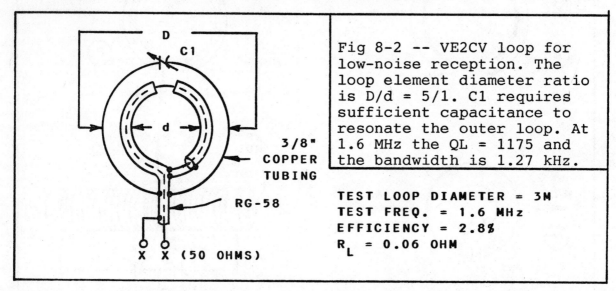

Fig 8-2 -- VE2CV loop for low-noise reception. The loop element diameter ratio is D/d = 5/1. C1 requires sufficient capacitance to resonate the outer loop. At 1.6 MHz the QL = 1175 and the bandwidth is 1.27 kHz.

TEST LOOP DIAMETER = 3M
TEST FREQ. = 1.6 MHz
EFFICIENCY = 2.8%
R_L = 0.06 OHM

My interest in the VE2CV loop was related to transmitting a 100-mW signal in the standard BC band (Part 15 of the FCC rules). The field strength at 1 Km was only 0.38 mV, which explains why small loops are not suitable for transmitting. The antenna of Fig 8-2 does, however, work very well for receiving if a low-noise preamplifier is used between it and the receiver. The nominal capacitance for resonance at 1.6 MHz with a 3-m outer loop diameter is approximately 735 pF.

On-Ground Wire Antennas

Some amateurs have experimented with "snake" antennas for low-noise reception. Many feet of RG-58 coaxial are laid on the ground in a straight line. The far end of the cable has the inner and outer conductors shorted, and a preamplifier is used between the receiver and the end nearest to the station. If you use a wavelength or more of coaxial cable for this antenna, it will probably work as a low-noise receiving antenna, however expensive this may be at 160 or 80 meters. I tried this antenna on 80 meters, and found that it was a fairly quiet device. I obtained better results when I terminated the far end of

the coaxial line with a 51-ohm resistor. This technique makes
it possible to use shorter lengths of cable, but at a sacrifice
in received-signal strength. A half wavelength of RG-58, shorted
at the far end, results in a short circuit at the receiver
end! Be sure to avoid multiples of half waves, inclusive of
the cable velocity factor (0.66), if you want to avoid having
the short circuit repeated at the opposite end of the line.
The 51-ohm terminating resistor eliminates this problem.

I developed a better on-ground antenna for my use on 160 meters.
It consists of one wavelength of two-conductor speaker wire.
I assumed a velocity factor of 0.75 for this type of wire,
and cut the antenna accordingly. Fig 8-3B shows how this antenna
is constructed.

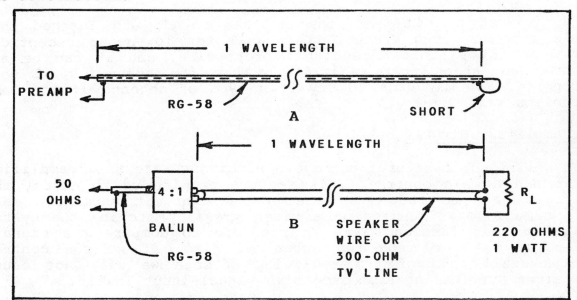

Fig 8-3 -- A coaxial "snake" antenna is seen at A.
One wavelength for 1.9 MHz is 341 feet, 10 inches.
For 3.6 MHz it is 180 feet, 5 inches. The two-wire
on-ground antenna at B offers improved performance.
No. 22 twin speaker cable is used for the antenna.
It is terminated at the far end by a 220-ohm resistor.
A 4:1 balun transformer converts the 200-ohm character-
istic to 50 ohms for feeding a preamplifier. Maximum
response for this antenna is off the terminated end.
It operates in the same manner as a Beverage antenna.

Perhaps you are wondering how these antennas can be less noisy
than a vertical or dipole. Good question! The mere fact that
we lay them on the ground does not provide the answer. Noise
fields in the area are less intense at ground level, compared
to the noise intensity near the main station antenna. Therefore,
our on-ground, long antenna picks up less man-made noise. Even
a random length of wire placed on the ground is often quieter
than the transmitting antenna. Also, the antenna at B of Fig
8-3 is directional. This means that noise fields off the sides

and receiver end of the antenna are rejected. This also minimizes noise pickup. Signals, on the other hand, are coming down from the sky, and impinge on the antenna, even though it is on the ground. Needless to say, the same antenna, if it were high above ground, would be more effective at collecting incoming-signal energy. The preamplifier compensates for the inefficiency of the on-ground antenna.

The antenna of Fig 8-3B would be 388 feet, 5 inches long. The same antenna for 3.6 MHz would be 205 feet long. Multiples of 1 wavelength (2, 3, etc.) provide greater directivity and permit increased signal pickup. Try to make your on-ground receiving antenna as long as practicable.

I have used a full-wave loop on the ground, and learned that this antenna works exceptionally well for low-noise reception. It does not exhibit directional properties, and it can be any convenient shape that will fit within the boundries of your property. You may want to try this type of antenna if you have problems with local QRN.

Preamplifier Circuit

You will find it helpful to have a receiving-antenna preamplifier that has a gain control. This feature permits us to adjust the circuit gain so that the receive and transmit antennas yield the same S-meter reading on a given signal. Matching the system gain in this fashion makes it possible to compare the signal-to-noise ratio of the two antennas. Also, if we can control the overall gain of the receiving antenna we will not cause receiver overloading from excessive signal-input levels.

When you use a separate receiving antenna it is necessary to protect the receiver input circuit from excessive RF current during the transmit period. It is not uncommon to find high values of RF voltage on the receive antenna when the transmitter is operating, especially when the receive antenna is resonant at the operating frequency. Use a TR (transmit-receive) relay to disconnect the receive antenna when the transmitter is operating. It is wise also to have an extra set of relay contacts that short circuit the receiver input line during transmit. If you use a preamplifier, place the relay protection circuit at the input of the preamplifier, since it then becomes your receiver-input circuit.

Fig 8-4 shows the circuit of a preamplifier that I have used for loops and on-ground receiving antennas. It works nicely with Beverage antennas as well. A grounded-gate JFET (Q1) is used as a low-noise preamplifier. A 2N4416 is suitable as a substitute device. C1 and C2 are trimmers that tune T1 and T2 to resonance at the operating frequency. A two-section variable capacitor may be substituted to ease peaking from the front panel of the preamplifier.

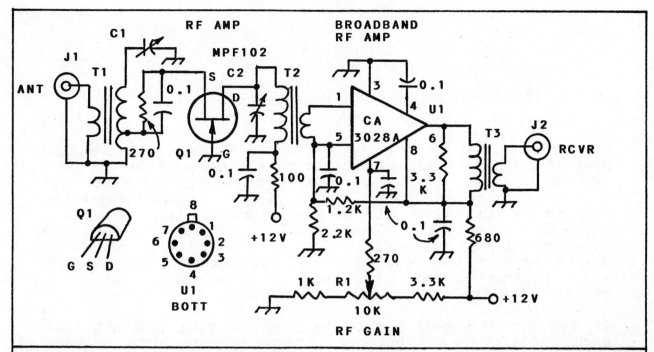

Fig 8-4 -- Schematic diagram of a 40-dB preamplifier that may be used from 1.8 to 30 MHz by selecting the correct values for C1, C2, T1 and T2. Capacitance is in µF and resistance is in ohms. K = 1000. Fixed-value capacitors are disc ceramic and fixed-value resistors are 1/4-W carbon comosition. C1 and C2 are 300-pF compression trimmers or a two-gang 365-pF variable capacitor. T1 has 23 turns of no. 28 enam. wire on an Amidon FT-37-61 ferrite core (30 µH) for 1.8 MHz. Tap at 6 turns from ground. Input link has 3 turns of no. 28 wire. For 3.5 MHz use 14 turns of no. 26 enam. wire (11 µH). Tap at 3 turns from ground. Input link has 2 turns. T2 is the same as T1 (minus tap). Output link has 6 turns for 1.8 MHz and 4 turns for 3.5 MHz. T3 has 14 primary turns of no. 26 enam. wire on an FT-37-43 ferrite toroid. Output link has 8 turns. R1 is a linear-taper carbon control. U1 is an RCA integrated circuit or equiv.

You may choose other ICs for U1, such as the Motorola MC1550G. A number of RF-amplifier ICs are suitable for this preamp if you make the appropriate circuit changes. Pin no. 2 of the IC specified in Fig 8-4 is not used. Keep all signal leads short, especially the gate lead of Q1. This will help to prevent circuit instability, which can be a problem when high gain is lumped in a small assembly.

Should you choose to use a two-section variable capacitor in place of C1 and C2, it will be necessary to connect a 3-30 pF trimmer in parallel with each section of the variable capacitor to ensure proper tracking of the tuned input circuit. The trimmer capacitors may be omitted if you use slug-tuned inductors for T1 and T2. The slugs will enable you to make the tuned circuits track.

Chapter 9

SIMPLE ANTENNA MEASUREMENTS

We need not have an expensive array of laboratory test equipment in order to evaluate our antennas. Some standard methods may be applied for this purpose, and we can build our own test gear at modest cost. After all, the primary objective we have in mind is to make the system radiate at its maximum capability. We must assume, of course, that the feed line has been matched to the transmitter, and that a proper match exists between the feed line and the antenna (center- and end-fed Zepps excluded). Maximum radiation occurs when the current flowing in the antenna is maximum.

It was not too many years ago that we hams did not have such gadgets as SWR bridges. A new amateur might wonder, "How in the world did those old timers survive without an SWR indicator or RF power meter?" The fact of the matter is that we did quite well! Perhaps the SWR on the line was not 1:1, or even 1.5:1. It may have been 2:1 or some other ratio, but we were heard in a decisive manner.

A common practice in the past was to insert an RF ammeter in series with the feed line (near the transmitter), then adjust the transmitter and antenna-matching network for the highest reading obtainable on the RF ammeter. Some hams used light bulbs in series with the feed line, then shorted across them after tuneup was completed. The idea was to tune for maximum bulb brilliance. This indicated maximum current flow to the antenna.

Many of us used (and still do) a field-strength indicator to measure the intensity of radiated waves from our antennas. It is reasonable to conclude that maximum field strength occurs when maximum current is flowing in the antenna. The field-strength meter, when practicable, is located one wavelength or greater from the antenna. The FS (field-strength) meter pickup antenna should be placed high above ground, so as to be in the field of maximum radiation. Also, the FS-meter antenna needs to be of the same polarization as the antenna under test (vertical or horizontal). The indicating meter can be connected remotely to the main instrument. This permits us to observe the meter while making antenna adjustments.

FS meters were the standard bill of fare in the past for adjusting mobile loading coils. Maximum field strength was noted when the antenna became resonant. If the mobile antenna included a matching network or coil at the feed point, it was also adjusted for maximum FS indication.

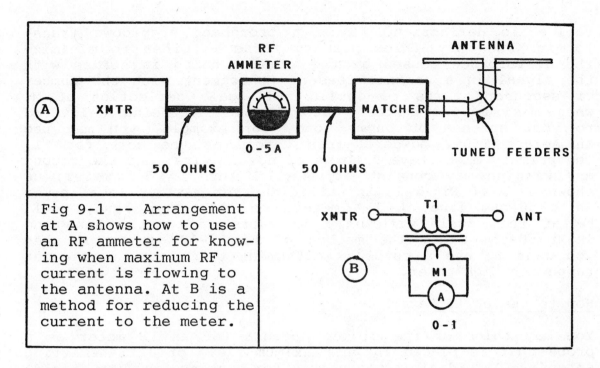

Fig 9-1 -- Arrangement at A shows how to use an RF ammeter for knowing when maximum RF current is flowing to the antenna. At B is a method for reducing the current to the meter.

The RF-Ammeter Method

Let's turn our attention toward the simple RF ammeter and how we may use it. Fig 9-1 shows the arrangement that I have used with a Transmatch and tuned feeders. In bygone days we amateurs sometimes used two RF ammeters, and placed them in each conductor of a balanced feed line, just after the Transmatch. This method worked, but the meter indication was sometimes quite low, owing to the feeder presenting a high impedance at some particular frequency. RF ammeters are intended to be used in low-impedance circuits. It is for this reason that I recommend they be placed in the 50-ohm line, as shown in Fig 9-1. When the impedance of the sampling point is known (50 ohms in this case), we can also calculate the transmitter output power. For example, if the meter reads 3 amperes in a 50-ohm line, the transmitter output power is 450 W ($W = I^2/R$). But, even if you do not wish to measure power, the ammeter is useful when tuning your Transmatch and transmitter for maximum power to the antenna. In other words, you may use it as a relative-reading instrument.

You need to be aware that RF ammeters are designed for use below 30 MHz, but I have seen some that were supposed to be accurate to 50 MHz. Also, they create an SWR "bump" in the transmission line. This is not significant at, say, 40 meters, but at 10 meters and higher the instrument causes an impedance discontinuity in a coaxial feed line. This can be checked by placing an SWR meter between the transmitter and the RF ammeter. The ammeter should be terminated by a 50-ohm dummy load for this test.

Fig 9-1B shows how to use an RF ammeter of lower current rating when the transmitter power is too great for normal

full-scale deflection. T1 is a broadband step-down transformer. Use only a few primary turns in order to minimize
the disturbance caused by the winding being in series with
the signal in a coaxial feeder. Experiment with the number
of secondary turns to obtain full-scale deflection at M1
when maximum transmitter output power is being delivered
to your antenna or dummy load. For example, you may use
an Amidon T80-2 powdered-iron toroid as the core for T1.
The primary might have 3 turns of no. 14 wire, and the secondary winding may consist of, say, 1 turn. A 5-A meter, as
shown at A of Fig 9-1, is sufficient for maximum legal power
in a 50-ohm line. A 1-A meter, on the other hand, would
be at full scale with only 50 watts of power, when used
in a 50-ohm line. The method of Fig 9-1B is useful when
you want to use a surplus instrument that lacks sufficient
range for high power.

How to Use an FS Meter

You may think of the pickup antenna for an FS meter as a
probe that is placed in the maximum field of a transmitting
antenna. An ideal arrangement for making quality field-
strength measurements is to employ a sensitive FS meter
that is located several wavelengths from the transmitting
antenna. There should be no conductive clutter in the path
between the antenna under test and the FS meter. In any
event, the FS measurements need to be made within the ground-
wave region of the radiation pattern.

The foregoing arrangement is generally outside the practical
reach of amateurs. We, because of limited real estate, are
forced to use our FS meters rather close to the antenna
under test. I recommend a distance of one wavelength or
greater. If this cannot be done, place your FS sampling
antenna as far from the transmitting antenna as possible.
Meaningful measurements will still be possible at less than
one wavelength of separation.

When we must conduct close-in measurements we should try
to elevate the FS sampling antenna to the height of the
test antenna. This is particularly important if the transmitting antenna has high-angle radiation. In other words, try
to place the FS-sampling antenna high enough to be in the
main lobe of the transmitting antenna. This is shown in
Fig 9-2. The FS-meter antenna need not be resonant at the
operating frequency. The larger it is, however, the greater
the sensitivity of the FS-meter system. I use a whip that
is approximately 30 inches long for the circuit in Fig 9-
3. I arrange to have the whip in a vertical position for
vertical antennas. I rotate the FS meter case by 90 degrees
to make the whip horizontal when testing horizontally polarized antennas. A short dipole may be substituted for the
whip.

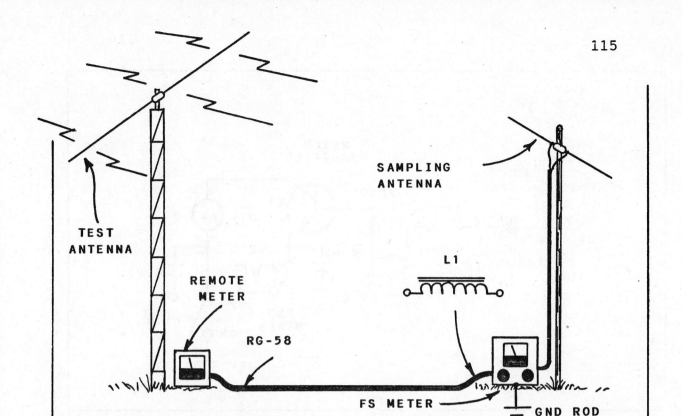

Fig 9-2 -- Arrangement for using a field-strength meter to monitor the results from adjusting a test antenna. L1 may be necessary to prevent the remote-meter cable from picking up unwanted RF energy and routing it into the FS meter (see text).

L1 of Fig 9-2 may be added to prevent unwanted RF energy from entering the FS meter via the remote-meter connecting line. RF current should reach the FS meter by way of the sampling antenna and its feed line, but not by way of other routes. If RF energy reaches the FS meter through the remote-meter line it can cause confusing meter readings. L1 consists of several close-wound turns of the RG-58 cable around an Amidon Assoc. 7 X 0.5-inch ferrite rod (850 mu). This unit acts as an RF choke. It should be placed close to the case of the FS meter.

You can place the remote meter near the supporting structure for your test antenna. Antenna-matching adjustments may then be carried out while noting the increase in the meter reading. This setup is useful also for observing the front-to-back and front-to-side radiation from a beam antenna. The power ratio between the front and back lobes can be determined if you calibrate the FS meter in dB. This may be done by adjusting the FS meter for a full-scale reading when your transmitter is delivering 50 or 100 watts to a dipole. Halve the transmitter power next, then note the reading on the FS meter. This represents a 3-dB decrease. Halve the power again and take a new meter reading. Continue this process until the FS meter reads zero. Use these 3-dB marks during tests of the beam-antenna pattern.

Fig 9-3 shows a circuit for a practical FS meter. A meter amplifier is included to permit the use of a 1-mA meter.

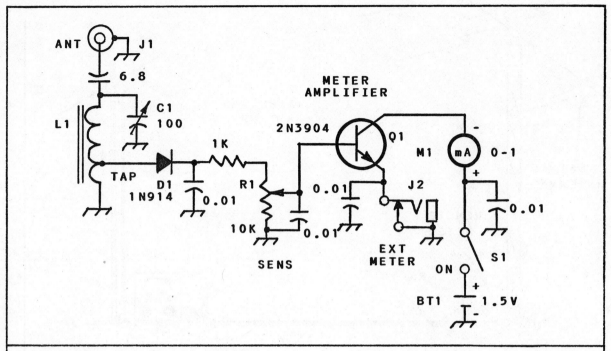

Fig 9-3 -- Schematic diagram of a practical FS meter. Decimal-value capacitors are in µF. Resistance is in ohms. K = 1000. BT1 is a AA cell. C1 is an air variable. L1 for 160 meters is 90 µH (36 ts no. 26 enam. wire on an Amidon FT-50-61 ferrite toroid. Tap at 8 ts from ground.). For 80/40 meters, use 19 ts of no. 24 enam. wire on an FT-50-61 toroid. Tap at 5 ts. For 30/20 meters (3 µH) use 25 ts. of no. 26 enam. wire on an Amidon T50-2 powdered-iron core. Tap at 6 ts. For 15/12/10 meters (0.8 µH) use 14 ts. of no. 26 enam. wire on a T50-6 toroid. Tap at 3 ts. M1 is a 0-1 mA dc meter (see text). Q1 may be any NPN small-signal transistor, such as a 2N2222 or 2N3904. R1 is a linear-taper carbon control.

The remote meter may be plugged in at J2. Use a 0-1 mA meter, as at M1 of Fig 9-3. Greater sensitivity may be had by using a 100- or 500-µA meter at M1. R1 is adjusted to provide a midscale reading when making FS measurements. This allows left- and right-changes in the meter reading for decreased or increased FS. This instrument may be used at 2 meters by substituting a 30-pF variable for C1. L1 will require 5 air-wound turns of no. 14 bus wire, 0.5 inch ID by 1 inch long. Tap at 1.5 turns from ground.

SWR Bridges

SWR indicators today follow the design of Warren Bruene of Collins Radio. A toroidal transformer is employed in a balanced bridge circuit. The transformer samples RF current and provides FORWARD and REFLECTED readings. This type of bridge is designed for the line impedance with which it will be used (50 or 75 ohms). The meter scale may be marked for SWR and/or RF power.

You will find an SWR meter useful in (1) learning the antenna resonant frequency and (2) adjusting the antenna-matching network to obtain an SWR of 1:1. Antenna resonance is noted at the frequency of lowest SWR as you vary the operating frequency within the band for which the antenna is designed. The observed dip in SWR, even though it may not be less than 2:1, indicates the resistive (nonreactive) frequency of the antenna (resonance). This assumes that you are making your measurements at the antenna feed point, or by means of a half wavelength of feed line (or multiple thereof). We must assume also that negligible harmonic current is present in the transmitter output, since harmonics of high magnitude can cause misleading SWR measurements.

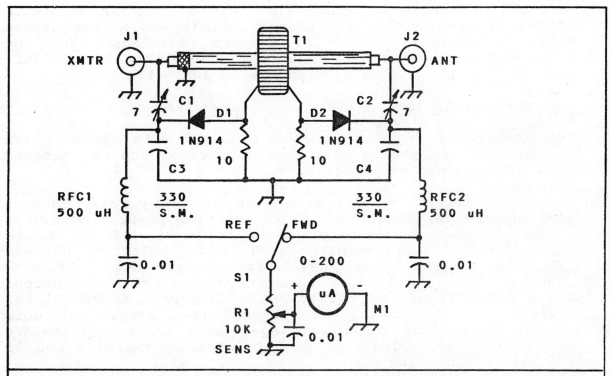

Fig 9-4 -- Schematic diagram of an SWR bridge that will operate from 1.8 through 50 MHz. Capacitance is in pF except for decimal-value units, which are in µF. Resistance is in ohms. K = 1000. C1 and C2 are small trimmers that have low minimum capacitance. Do not use mica trimmers. C3 and C4 are matched silver-mica units. M1 may be any microammeter from 50 to 200 µA. R1 is a linear-taper carbon control. S1 may be any SPDT switch. T1 has 35 turns of no. 24 enam. wire on an Amidon T68-2 powdered-iron toroid (25 mu).

A Practical SWR Bridge

Fig 9-4 contains the circuit for a simple SWR bridge that you can use from 160 through 6 meters. A short length of 50-ohm coaxial cable is passed through the toroid core to form T1. The coaxial line is the transformer primary, and is equivalent to one turn. The shield braid of the cable is grounded only at the J1 end. This method permits the cable outer conductor to function as a Faraday screen. This blocks the passage of harmonic currents to the bridge circuit.

The two 10-ohm resistors should be matched in value. They are 1/2-W, 5% units. Trimmers C1 and C2 must have a very low minimum capacitance in order to provide ample range for balancing the bridge. Glass piston trimmers are suitable, and so are miniature air trimmers. Most ceramic and plastic trimmers have too high a minimum capacitance to work in this circuit. Likewise with mica compression trimmers.

You may use any μA dc meter that is available. Greatest sensitivity will occur if you use a 50-μA instrument (this is a good value to use for low-power operation). The bridge sensitivity may be increased also by using 1N34A germanium diodes at D1 and D2. In either event, try to select diodes that are closely matched in forward resistance. You may check this by using a VOM. Silicon diodes (1N914 or similar) are best for power levels in excess of 100 watts.

Calibration and Use

The circuit of Fig 9-4 must be laid out with symmetry and short leads. If you have accomplished this, you may proceed with checkout and calibration.

Terminate the bridge (J2) with a 50-ohm dummy load. Connect your transmitter to J1. Set S1 to read FWD power, and adjust R1 for minimum sensitivity. Apply RF power and set R1 to provide a full-scale reading on M1. Next, switch S1 to read REF power. Ideally, the meter should read zero. If not, adjust C2 for a null in REF power. Reverse the connections for J1 and J2, and set S1 to read FWD power. Adjust C1 for a null in meter indication. Repeat these steps once more for final nulling of the bridge. Make these adjustments at 14 or 21 MHz to ensure good balance at the high end of the bridge operating range.

If you build your SWR bridge on a PC board, be sure to use single-sided material (copper on only one side). Try to maintain a spacing of at least 1/8 inch between PC-board foils. This will minimize unwanted stray capacitance, which can prevent you from balancing the bridge.

Antenna or Transmatch adjustments are done by tuning for minimum SWR with the bridge between the transmitter and feed line, or between the transmitter and the Transmatch. A zero REF reading indicates an SWR of 1:1.

Complete construction details for this and a resistive SWR bridge for QRP are given in "The SWR Twins -- QRP and QRO," **QST** for July 1986, page 34. Also, the measurements chapter of recent editions of **The ARRL Handbook** provides construction details for SWR bridges.

Antenna Noise Bridge

Antenna evaluation and adjustment becomes more of a science when we use noise bridges for measuring the resistance and reactance values of a system. The SWR bridge of Fig 9-4 lets us know the resonant frequency of an antenna (the dip point in the SWR reading), but it cannot tell us what the resistance is in ohms. A noise bridge, on the other hand, indicates the resistance on a calibrated scale. Similarly, we may read directly the reactance of the system (capacitive or inductive).

The noise-bridge reactance scale is at zero when the load attached to the bridge is purely resistive. The minus part of the reactance scale indicates that Xc (capacitive reactance) is present. If the bridge nulls when the reactance scale is in the plus range, there is XL (inductive reactance). When we make antenna measurements with a noise bridge we know that the antenna is too short if the null occurs with an Xc reading on the dial. Similarly, an XL-setting null tells us that the antenna is too long. Fig 9-5 shows the circuit for a noise bridge.

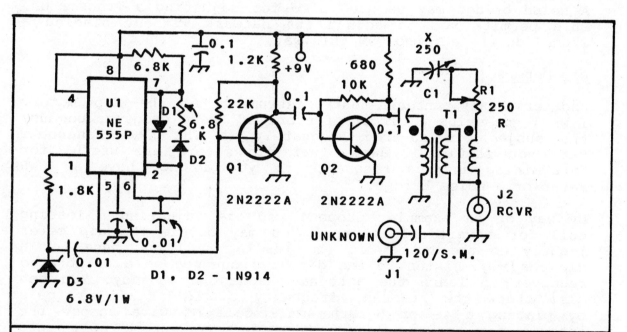

Fig 9-5 -- Schematic diagram of a noise bridge. Decimal-value capacitors are in μF. Others are pF. The black dots at T1 indicate winding polarity. C1 is a 250-pF air variable, with low minimum capacitance. R1 should be a high-quality Allen-Bradley type of linear-taper carbon control. Keep all leads short, especially in T1 area. T1 is trifilar wound, 8 turns of no. 26 enam. wire on an Amidon FT-37-43 toroid core (850 mu).

A detailed discussion of this circuit is provided in the
measurements chapter of **The ARRL Handbook.** That section
also includes a PC-board pattern and a parts-placement guide.

The circuit of Fig 9-5 contains an audio generator (U1)
that produces a 1000-Hz square-wave signal with a 50% duty
cycle. This permits exact nulling of the bridge when it
is used with an AM detector. As the nulling process continues
(C1 and R1 adjustment), the 1000-Hz tone becomes more and
more audible over the wide-band noise that is generated
by Zener diode D3. The noise becomes lower and lower in
amplitude as the null is reached, but the audio tone sounds
louder.

Your noise bridge should be used at the antenna feed point
when practicable. If this is not easy for you to do, connect
it to the antenna through a half wavelength of coaxial line,
inclusive of the velocity factor of the cable. This cable
will repeat the impedance at the antenna feed point. You
may use multiples of 0.5 wavelength if necessary. The cable
length may be calculated from L(ft) = 492/f(MHz) x VF, where
VF is the velocity factor of the feed line used.

A noise bridge may be used also for adjusting a Transmatch.
This permits you to adjust the network without placing a
signal on the air (no QRM this way).

Dip Meters

Another handy antenna-test instrument is the dip meter.
I won't describe a circuit for one of these gadgets, because
the subject is covered adequately in **The ARRL Handbook.**
But, you should be aware that dip meters are useful for
determining antenna resonance if you do not have an SWR
meter or a noise bridge.

By way of an example, suppose you are adjusting a loading
coil for a vertical antenna. You may couple the dip meter
lightly to the bottom of the loading coil and tune for a
dip reading. Listen to the dipper frequency on a calibrated
receiver to learn the antenna resonant frequency. You can
also check the resonant frequency of full-size verticals
by placing a 4- or 5-turn small coil of wire across the
antenna feed point. Insert the dip-meter probe coil into
this loop and find the dip frequency.

A small coil across the feeder to a beam antenna will allow
you to check the resonant frequency of the antenna elements.
For example, there will be a dip for the director, driven
element and the reflector. An additional dip will probably
show up for the resonant frequency of the feed line.

The length of a section of feed line may also be checked
for resonance by means of a dip meter. Terminate the line

in its characteristic impedance (50 or 75 ohms, for example). Connect a small sampling loop (4 or 5 turns) across the conductors at the opposite end of the line. Insert the dipper coil into the sampling loop and tune for a dip.

Dip meters are invaluable for adjusting antenna traps to resonance. The traps need not be connected in the antenna system when they are tuned to the desired frequency.

Current Sampling for Verticals

A true measure of radiation efficiency for an antenna is related to the magnitude of the current that flows in the radiator. I have used the system of Fig 9-6 to sample the current in my verticals. This gave me a proper indication that my matching network was adjusted for maximum power transfer. It also provided an indication of how well the ground system was working as I added radial wires.

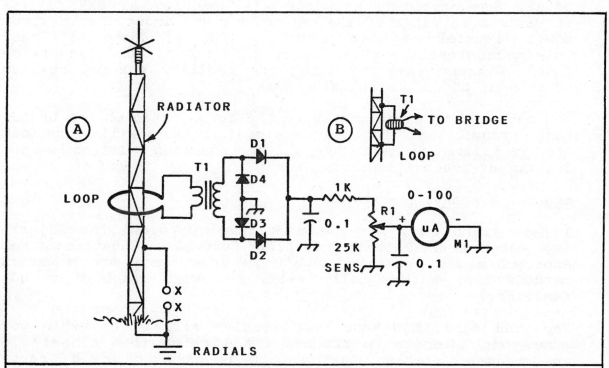

Fig 9-6 -- Current sampling methods for use with vertical antennas. Method A calls for a single-turn loop of no. 14 insulated wire around the tower or mast. The loop is connected to broadband transformer T1. A bridge rectifier (1N914 diodes) changes the RF current to dc. M1 reads the voltage from the rectifiers. M1 is a 100- or 200-µA dc meter. R1 is a linear-taper carbon control. T1 (for under 150 W) has a 2-turn primary and a 6-turn secondary. For QRO use 2 turns for each winding. See text for further information.

Two current-sampling methods are illustrated in Fig 9-6. The technique at A calls for a single-turn loop of wire around the radiator. It is attached to broadband transformer T1. The transformer uses no. 24 enamel wire that is wound on an FT-50-43 ferrite toroid (850 mu). See Fig 9-6 for winding information. T1 and the remainder of the detector circuit may be used at ground level. Connect a suitable length of RG-58 coaxial cable to the sampling loop and route it to ground. Tape the RG-58 line to the tower leg. Ground the shield braid at the base of the tower.

Fig 9-6B shows an alternative method for sampling the current in a vertical antenna. A 1-foot-long rod (1/8 inch OD) is attached to the tower as shown. It passes through the toroid core and becomes the primary for the transformer. You may use the same type of core specified at A of Fig 9-6. The secondary winding requires 15 turns of no. 26 enamel wire. The sampling loops at A and B of Fig 9-6 are placed 10 or 15 feet above ground on the radiator. These methods are superior to the SWR-bridge technique, especially when lumped-constant LC matching networks are used at the antenna feed point. Some matching networks will provide an SWR of 1:1 at various settings of the capacitors and inductors. Maximum power transfer may not occur at some of these settings. This means that power is lost within the network (insertion loss). Current sampling along the radiator ensures optimum adjustment of the matching network.

You may want to experiment with the turns ratio of the broad-band transformer. Too much signal pickup will burn out the rectifier diodes. Too little coupling will make the instrument insensitive.

Chapter Summary

I have tried to cover the more common devices for adjusting antennas and their matching networks. Countless other approaches are possible, and I am sure there are numerous methods that work equally well, but with which I am not familiar.

You will find that your best results will be had when you tune your antennas to resonance and match them closely to the transmission line. Although it is easy to use a Transmatch to disguise a high SWR condition, try to avoid that trap. The antenna system will deliver far better results when the impedances are matched and the SWR is low.

INDEX

Schematic Symbols Used in Circuit Diagrams

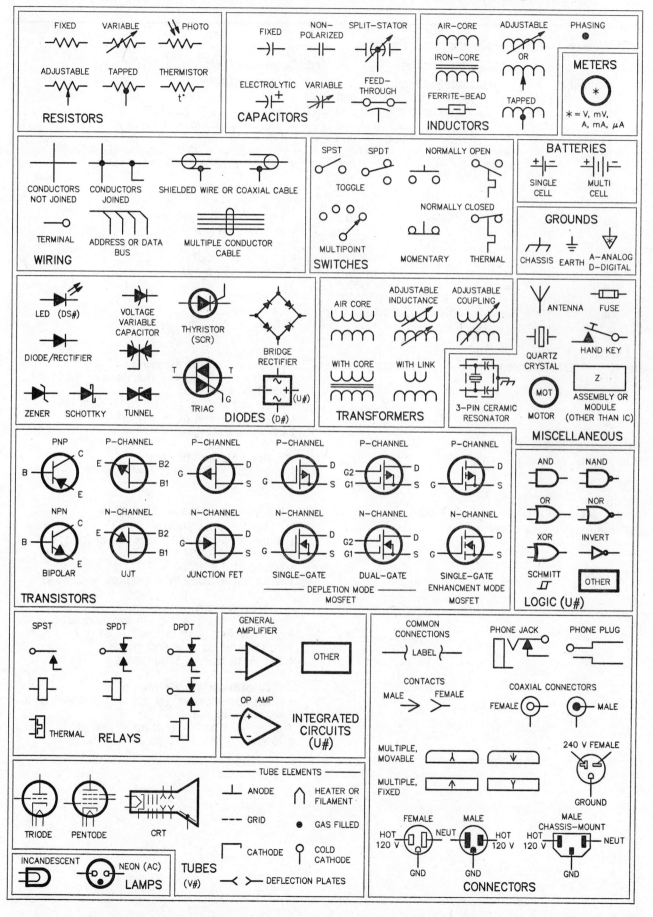

U.S. Customary — Metric Conversion Factors

International System of Units (SI) — Metric Units

Prefix	Symbol		Multiplication Factor
exa	E	10^{18} =	1,000,000,000,000,000,000
peta	P	10^{15} =	1,000,000,000,000,000
tera	T	10^{12} =	1,000,000,000,000
giga	G	10^{9} =	1,000,000,000
mega	M	10^{6} =	1,000,000
kilo	k	10^{3} =	1,000
hecto	h	10^{2} =	100
deca	da	10^{1} =	10
(unit)		10^{0} =	1
deci	d	10^{-1} =	0.1
centi	c	10^{-2} =	0.01
milli	m	10^{-3} =	0.001
micro	μ	10^{-6} =	0.000001
nano	n	10^{-9} =	0.000000001
pico	p	10^{-12} =	0.000000000001
femto	f	10^{-15} =	0.000000000000001
atto	a	10^{-18} =	0.000000000000000001

Linear
1 meter (m) = 100 centimeters (cm) = 1000 millimeters (mm)

Area
$1 \text{ m}^2 = 1 \times 10^4 \text{ cm}^2 = 1 \times 10^6 \text{ mm}^2$

Volume
$1 \text{ m}^3 = 1 \times 10^6 \text{ cm}^3 = 1 \times 10^9 \text{ mm}^3$
$1 \text{ liter (l)} = 1000 \text{ cm}^3 = 1 \times 10^6 \text{ mm}^3$

Mass
1 kilogram (kg) = 1000 grams (g)
  (Approximately the mass of 1 liter of water)
1 metric ton (or tonne) = 1000 kg

U.S. Customary Units

Linear Units
12 inches (in) = 1 foot (ft)
36 inches = 3 feet = 1 yard (yd)
1 rod = 5½ yards = 16½ feet
1 statute mile = 1760 yards = 5280 feet
1 nautical mile = 6076.11549 feet

Area
$1 \text{ ft}^2 = 144 \text{ in}^2$
$1 \text{ yd}^2 = 9 \text{ ft}^2 = 1296 \text{ in}^2$
$1 \text{ rod}^2 = 30¼ \text{ yd}^2$
$1 \text{ acre} = 4840 \text{ yd}^2 = 43{,}560 \text{ ft}^2$
$1 \text{ acre} = 160 \text{ rod}^2$
$1 \text{ mile}^2 = 640 \text{ acres}$

Volume
$1 \text{ ft}^3 = 1728 \text{ in}^3$
$1 \text{ yd}^3 = 27 \text{ ft}^3$

Liquid Volume Measure
$1 \text{ fluid ounce (fl oz)} = 8 \text{ fluidrams} = 1.804 \text{ in}^3$
1 pint (pt) = 16 fl oz
$1 \text{ quart (qt)} = 2 \text{ pt} = 32 \text{ fl oz} = 57¾ \text{ in}^3$
$1 \text{ gallon (gal)} = 4 \text{ qt} = 231 \text{ in}^3$
1 barrel = 31½ gal

Dry Volume Measure
$1 \text{ quart (qt)} = 2 \text{ pints (pt)} = 67.2 \text{ in}^3$
1 peck = 8 qt

$1 \text{ bushel} = 4 \text{ pecks} = 2150.42 \text{ in}^3$

Avoirdupois Weight
1 dram (dr) = 27.343 grains (gr) or (gr a)
1 ounce (oz) = 437.5 gr
1 pound (lb) = 16 oz = 7000 gr
1 short ton = 2000 lb, 1 long ton = 2240 lb

Troy Weight
1 grain troy (gr t) = 1 grain avoirdupois
1 pennyweight (dwt) or (pwt) = 24 gr t
1 ounce troy (oz t) = 480 grains
1 lb t = 12 oz t = 5760 grains

Apothecaries' Weight
1 grain apothecaries' (gr ap) = 1 gr t = 1 gr a
1 dram ap (dr ap) = 60 gr
1 oz ap = 1 oz t = 8 dr ap = 480 gr
1 lb ap = 1 lb t = 12 oz ap = 5760 gr

Multiply ⟶

Metric Unit = Conversion Factor × U.S. Customary Unit

⟵ Divide

Metric Unit ÷ Conversion Factor = U.S. Customary Unit

Metric Unit	=	Conversion Factor ×	U.S. Unit	Metric Unit	=	Conversion Factor ×	U.S. Unit
(Length)				**(Volume)**			
mm		25.4	inch	mm^3		16387.064	in^3
cm		2.54	inch	cm^3		16.387	in^3
cm		30.48	foot	m^3		0.028316	ft^3
m		0.3048	foot	m^3		0.764555	yd^3
m		0.9144	yard	ml		16.387	in^3
km		1.609	mile	ml		29.57	fl oz
km		1.852	nautical mile	ml		473	pint
(Area)				ml		946.333	quart
mm^2		645.16	$inch^2$	l		28.32	ft^3
cm^2		6.4516	in^2	l		0.9463	quart
cm^2		929.03	ft^2	l		3.785	gallon
m^2		0.0929	ft^2	l		1.101	dry quart
cm^2		8361.3	yd^2	l		8.809	peck
m^2		0.83613	yd^2	l		35.238	bushel
m^2		4047	acre				
km^2		2.59	mi^2				
(Mass)		**(Avoirdupois Weight)**		**(Mass)**		**(Troy Weight)**	
grams		0.0648	grains	g		31.103	oz t
g		28.349	oz	g		373.248	lb t
g		453.59	lb	**(Mass)**		**(Apothecaries' Weight)**	
kg		0.45359	lb	g		3.387	dr ap
tonne		0.907	short ton	g		31.103	oz ap
tonne		1.016	long ton	g		373.248	lb ap

Glossary of Antenna Terms

Actual Ground — The point within the earth's surface where effective ground conductivity exists. The depth for this point varies with frequency, the condition of the soil, and the geographical region.

Antenna — An electrical conductor or array of conductors that radiates signal energy (transmitting) or collects signal energy (receiving).

Aperture, Effective — An area enclosing an antenna, on which it is convenient to make calculations of field strength and antenna gain. Sometimes referred to as the "capture area."

Apex — The feed-point region of a V type of antenna.

Apex Angle — The enclosed angle in degrees between the wires of a V, an inverted V, and similar antennas.

Balanced Line — A symmetrical two-conductor feed line that can have uniform voltage and current distribution along its length.

Balun — A device for feeding a balanced load with an unbalanced line, or vice versa. May be a form of choke, or a transformer that provides a specific impedance transformation (including 1:1). Often used in antenna systems to interface a coaxial transmission line to the feed point of a balanced antenna, such as a dipole.

Base Loading — A coil of specific reactance value that is inserted at the base (ground end) of a vertical antenna to cancel capacitive reactance and resonate the antenna.

Bazooka — A transmission-line balancer. It is a quarter-wave conductive sleeve (tubing or flexible shielding) placed at the feed point of a center-fed dipole and grounded to the shield braid of the coaxial feed line at the end of the sleeve farthest from the feed point. It permits the use of unbalanced feed line with balanced-feed antennas.

Beamwidth — Related to directive antennas. The width, in degrees, of the major lobe between the two directions at which the relative radiated power is equal to one half its value (half-power or −3-dB points) at the peak of the lobe.

Beta Match — Sometimes called a "hairpin match." It is a U-shaped conductor that is connected to the two inner ends of a split dipole for the purpose of creating an impedance match to a balanced feeder.

Bridge — A circuit with two or more ports that is used in measurements of impedance, resistance or standing waves in an antenna system. When the bridge is adjusted for a balanced condition, the unknown factor can be determined by reading its value on a calibrated scale or meter.

Capacitance Hat — A conductor of large surface area that is connected to an antenna to lower its resonant frequency. It is sometimes mounted directly above a loading coil to reduce the required inductance for establishing resonance. It usually takes the form of a series of wheel spokes or a solid circular disc. Sometimes referred to as a "top hat."

Capture Area — See aperture.

Center Fed — Transmission-line connection at the electrical center of an antenna radiator.

Center Loading — A scheme for inserting inductive reactance (coil) at or near the center of an antenna element for the purpose of resonating it. Used with elements that are less than 1/4 wavelength

Coax Cable — Any of the coaxial transmission lines that have the outer shield (solid or braided) on the same axis as the inner or center conductor. The insulating material can be air, helium or solid-dielectric compounds.

Collinear Array — A linear array of radiating elements (usually dipoles) with their axes arranged in a straight line. Popular at vhf and higher.

Conductor — A metal body such as tubing, rod or wire that permits current to travel continuously along its length.

Counterpoise — A wire or group of wires mounted close to ground, but insulated from ground, to form a low-impedance, high-capacitance path to ground. Used at mf and hf to provide an effective ground for an antenna.

Current Loop — The point of maximum current (antinode) on an antenna. The minimum point is called a "node."

Delta Loop — A full-wave loop shaped like a triangle or delta.

Delta Match — Matching technique used with half-wave radiators that are not split at the center. The feed line is fanned near the radiator center and connected to the radiator symmetrically. The fanned area is delta-shaped.

Dielectrics — Various insulating materials used in antenna systems, such as found in insulators and transmission lines.

Dipole — An antenna that is split at the exact center for connection to a feed line. Usually a half wavelength in dimension. Also called a "doublet."

Directivity — The property of an antenna that concentrates the radiated energy to form one or more major lobes.

Director — A conductor placed in front of a driven element to cause directivity. Frequently used singly or in multiples with Yagi or cubical-quad beam antennas.

Direct Ray — Transmitted signal energy that arrives at the receiving antenna directly rather than being reflected from the ionosphere, ground or a man-made passive reflector.

Doublet — See dipole.

Driven Array — An array of antenna elements which are all driven or excited by means of a transmission line.

Driven Element — The radiator element of an antenna system. The element to which the transmission line is connected.

E Layer — The ionospheric layer nearest earth from which radio signals can be reflected to a distant point, generally a maximum of 2000 km (1250 mi.).

E Plane — The plane containing the axis of the antenna and the electric field vector of an antenna.

Efficiency — The ratio of useful output power to input power, determined in antenna systems by losses in the system, including in nearby objects.

Elements — The conductive parts of an antenna system that determine the antenna characteristics. For example, the reflector, driven element and directors of a Yagi antenna.

End Effect — A condition caused by capacitance at the ends of an antenna element. Insulators and related support wires contribute to this capacitance and effectively lower the resonant frequency of the antenna. The effect increases with diameter and must be considered when cutting an antenna element to length.

End Fed — An end-fed antenna is one to which power is applied at one end rather than at some point between the ends.

F Layer — The ionospheric layer that lies above the E layer. Radio waves can be reflected from it to provide communications ranges of several thousand miles by means of single- or double-hop skip.

Feeders — Transmission lines of assorted types that are used to route rf power from a transmitter to an antenna, or from an antenna to a receiver.

Feed Line — See feeders.

Field Strength — The intensity of a radio wave as measured at a point some distance from the antenna. This measurement is usually made in terms of microvolts per meter.

Front to Back — The ratio of the radiated power off the front and back of a directive antenna. A dipole would have a ratio of 1, for example.

Front to Side — The ratio of radiated power between the major lobe and the null side of a directive antenna.

Gain — Increase in effective radiated power in the desired direction of the major lobe.

Gamma Match — A matching system used with driven antenna elements to effect a match between the transmission line and the feed point of the antenna. It consists of an adjustable arm that is mounted close to the driven element and in parallel with it near the feed point.

Ground Plane — A man-made system of conductors placed below an antenna to serve as an earth ground.

Ground Screen — A wire mesh ground plane.

Ground Wave — Radio waves that travel along the earth's surface to the receiving point.

H Plane — Related to a linearly polarized antenna. The plane that is perpendicular to the axis of the elements and contains the magnetic field vector.

Harmonic Antenna — An antenna that will operate on its fundamental frequency and the harmonics of the fundamental frequency for which it is designed. An end-fed half-wave antenna is one example.

Helical — A helical or helically wound antenna is one that consists of a spiral conductor. If it has a very large winding length to diameter ratio it provides broadside radiation. If the length/diameter ratio is small, it will operate in the axial mode and radiate off the end opposite the feed point. The polarization will be circular for the axial mode, with left or right circularity, depending on whether the helix is wound clockwise or counter-clockwise.

Image Antenna — The imaginary counterpart of an actual antenna. It is assumed for mathematical purposes to be located below the earth's surface beneath the antenna, and is considered symmetrical with the antenna above ground.

Impedance — The ohmic value of an antenna feed point, matching section or transmission line. An impedance may contain a reactance as well as a resistance component.

Inverted V — A half-wavelength dipole erected in the form of an upside-down V, with the feed point at the apex. It is essentially omnidirectional, and is sometimes called a "drooping doublet."

Isotropic — An imaginary or hypothetical antenna in free space that radiates equally in all directions. It is used as a reference for the directive characteristics of actual antennas.

Lambda — Greek symbol (λ) used to represent a wavelength with reference to electrical dimensions in antenna work.

Line Loss — The power lost in a transmission line, usually expressed in decibels.

Line of Sight — Transmission path of a wave that travels directly from the transmitting antenna to the receiving antenna.

Load — The electrical entity to which power is delivered. The antenna is a load for the transmitter. A dummy load is a nonradiating substitute for an antenna.

Loading — The process of a transferring power from its source to a load. The effect a load has on a power source.

Lobe — A defined field of energy that radiates from a directive antenna.

Log Periodic Antenna — A broadband directive antenna that has a structural format which causes its impedance and radiation characteristics to repeat periodically as the logarithm of frequency.

Long Wire — A wire antenna that is one wavelength or greater in electrical length. When two or more wavelengths long it provides gain and a multilobe radiation pattern. When terminated at one end it becomes essentially unidirectional off that end.

Marconi Antenna — Any type of vertical monopole operated against ground or a radial system.

Matching — The process of effecting an impedance match between two electrical circuits of unlike impedance. One example is matching a transmission line to the feed point of an antenna. Maximum power transfer to the load (antenna system) will occur when a matched condition exists.

Null — A condition during which an electrical property is at a minimum. The null in an antenna radiation pattern is that point in the 360-degree pattern where minimum field intensity is observed. An impedance bridge is said to be "nulled" when it has been brought into balance.

Open-Wire Line — A type of transmission line that resembles a ladder, sometimes called "ladder line." Consists of parallel, symmetrical wires with insulating spacers every few inches to maintain the line spacing. The dielectric is principally air, making it a low-pass type of line.

Parabolic Reflector — An antenna reflector that is a portion of a parabolic revolution or curve. Used mainly at uhf and higher to obtain high gain and a relatively narrow beamwidth when excited by one of a variety of driven elements placed in the plane of and perpendicular to the axis of the parabola.

Parasitic Array — A directive antenna that has a driven element and independent directors, a reflector, or both. The directors and reflector are not connected to the feed line. A Yagi antenna is one example. See Driven Array.

Phasing Lines — Sections of transmission line that are used ensure correct phase relationship between the bays of an array of antennas. Also used to effect impedance transformations while maintaining the desired array phase.

Polarization — The polarization of the wave radiated by an antenna. This can be horizontal, vertical, elliptical or circular (left- or right-hand circularity), depending on the design and application.

Q Section — Term used in reference to transmission-line matching transformers and phasing lines.

Quad — Rectangular or diamond-shaped full-wave wire-loop antenna. Most often used with a parasitic loop director and a parasitic loop reflector to provide approximately 8 dB of gain and good directivity. Often called the "cubical quad." Another version uses delta-shaped elements, and is called a "delta loop" beam.

Random Wire — A random length of wire used as an antenna and fed at one end by means of a Transmatch. Seldom operates as a resonant antenna unless the length happens to be correct.

Radiation Pattern — The radiation characteristics of an antenna as a function of space coordinates. Normally, the pattern is measured in the far-field region and is represented graphically.

Radiation Resistance — The ratio of the power radiated by an antenna to the square of the rms antenna current, referred to a specific point and assuming no losses. The effective resistance at the antenna feed point.

Radiator — A discrete conductor in an antenna system that radiates rf energy. The element to which the feed line is attached.

Reflected Ray — A radio wave that is reflected from earth, the ionosphere or a man-made medium, such as a passive reflector.

Reflector — A parasitic antenna element or a metal assembly that is located behind the driven element to enhance forward directivity. Hillsides and large manmade structures such as buildings and towers may reflect radio signals.

Refraction — Process by which a radio wave is bent and returned to earth from an ionospheric layer or other medium after striking the medium.

Rhombic — A rhomboid or diamond-shaped antenna consisting of sides (legs) that are each one wavelength or greater in electrical length. The conductors are made from wire, and the antenna is usually erected parallel to the ground. A rhombic antenna is bidirectional unless terminated by a resistance, at which time it is predominently unidirectional. The greater the leg length, the greater the gain.

Shunt Feed — A method of feeding an antenna driven element with a parallel conductor mounted adjacent to a low-impedance point on the radiator. Frequently used with grounded quarter-wave vertical antennas (Marconis) to provide an impedance match to the feeder. Series feed is used when the base of the vertical is insulated from ground.

Source — The point of origination (transmitter or generator) for rf power supplied to an antenna system.

Stacking — The process of placing similar directive antennas atop or beside one another, forming a "stacked array."

Stub — A section of transmission line used to tune an antenna element to resonance or to aid in obtaining an impedance match.

SWR — Standing-wave ratio on a transmission line in an antenna system. More correctly, "VSWR," or *voltage standing wave ratio*. The ratio of the forward to reflected voltage on the line, and not a power ratio. A VSWR of 1:1 occurs when all parts of the antenna system are matched correctly to one another.

Tilt Angle — Half the angle included between the wires at the sides of a rhombic antenna.

T Match — Method for matching a transmission-line to an unbroken driven element. Attached at the electrical center of the driven element in a T-shaped manner. In effect it is a double gamma match.

Top Hat — Capacitance hat used at the high-impedance end of a quarter-wave driven element to effectively increase the electrical length. See Capacitance Hat.

Top Loading — Addition of inductive reactance (coil) and/or a capacitance hat at the end of a driven element opposite the feed point to increase the electrical length of the radiator.

Traps — Parallel L-C networks inserted in an antenna element to provide multiband operation with a single conductor.

Velocity Factor — That which affects the speed of radio waves in accordance with the dielectric medium they are in. A factor of 1 is applied to the speed of light and radio waves in free space, but the velocity is reduced in various dielectric mediums, such as transmission lines. When cutting a transmission line to a specific electrical length, the velocity factor of the particular line must be taken into account.

VSWR — Voltage standing-wave ratio. See SWR.

Wave — A disturbance that is a function of time or space, or both. A radio wave, for example.

Wave Angle — The angle above the horizon of a radio wave as it is launched from an antenna.

Wave Front — A continuous surface that is a locus of points having the same phase at a specified instant.

Yagi — A directive, gain type of antenna that utilizes a number of parasitic directors and a reflector. Named after one of the inventors (Yagi and Uda).

Zepp Antenna — A half-wave wire antenna that operates on its fundamental and harmonics. Fed at one end (end-fed Zepp) or at the center (center-fed Zepp) by means of open-wire feeders. The name evolved from its popularity as an antenna on zeppelins.

The Decibel

The decibel and its use are discussed in Chapter 2. Table 2 below shows the number of decibels corresponding to various power and voltage ratios.

The decibel value is read from the body of the table for the desired ratio, including decimal increment. For example, a *power* ratio of 2.6 is equivalent to 4.15 dB. A *voltage* ratio of 4.3 (voltages measured across like impedances) is equivalent to 12.67 dB. Values from the table may be extended, as indicated at the lower left in each section. For example, a *power* ratio of 17, which is the same as 10 × 1.7, is equivalent to 10 + 2.30 = 12.30 dB. Similarly, a power ratio of 170 (100 × 1.7) = 20 + 2.30 = 22.30 dB.

Table 2
Power Ratio to Decibel Conversion

Ratio	Decimal Increments									
	0.0	0.1	0.2	0.3	0.4	0.5	0.6	0.7	0.8	0.9
1	0.00	0.41	0.79	1.14	1.46	1.76	2.04	2.30	2.55	2.79
2	3.01	3.22	3.42	3.62	3.80	3.98	4.15	4.31	4.47	4.62
3	4.77	4.91	5.05	5.19	5.32	5.44	5.56	5.68	5.80	5.91
4	6.02	6.13	6.23	6.34	6.44	6.53	6.63	6.72	6.81	6.90
5	6.99	7.08	7.16	7.24	7.32	7.40	7.48	7.56	7.63	7.71
6	7.78	7.85	7.92	7.99	8.06	8.13	8.20	8.26	8.33	8.39
7	8.45	8.51	8.57	8.63	8.69	8.75	8.81	8.86	8.92	8.98
8	9.03	9.08	9.14	9.19	9.24	9.29	9.34	9.40	9.44	9.49
9	9.54	9.59	9.64	9.68	9.73	9.78	9.82	9.87	9.91	9.96
10	10.00	10.04	10.09	10.13	10.17	10.21	10.25	10.29	10.33	10.37
×10	+10									
×100	+20									
×1000	+30									
×10,000	+40									
×100,000	+50									

Voltage Ratio to Decibel Conversion

Ratio	Decimal Increments									
	0.0	0.1	0.2	0.3	0.4	0.5	0.6	0.7	0.8	0.9
1	0.00	0.83	1.58	2.28	2.92	3.52	4.08	4.61	5.11	5.58
2	6.02	6.44	6.85	7.23	7.60	7.96	8.30	8.63	8.94	9.25
3	9.54	9.83	10.10	10.37	10.63	10.88	11.13	11.36	11.60	11.82
4	12.04	12.26	12.46	12.67	12.87	13.06	13.26	13.44	13.62	13.80
5	13.98	14.15	14.32	14.49	14.65	14.81	14.96	15.12	15.27	15.42
6	15.56	15.71	15.85	15.99	16.12	16.26	16.39	16.52	16.65	16.78
7	16.90	17.03	17.15	17.27	17.38	17.50	17.62	17.73	17.84	17.95
8	18.06	18.17	18.28	18.38	18.49	18.59	18.69	18.79	18.89	18.99
9	19.08	19.18	19.28	19.37	19.46	19.55	19.65	19.74	19.82	19.91
10	20.00	20.09	20.17	20.26	20.34	20.42	20.51	20.59	20.67	20.75
×10	+20									
×100	+40									
×1000	+60									
×10,000	+80									
×100,000	+100									

Length Conversions

Throughout this book, equations may be found for determining the design length and spacing of antenna elements. For convenience, the equations are written to yield a result in feet. (The answer may be converted to meters simply by multiplying the result by 0.3048.) If the result in feet is not an integral number, however, it is necessary to make a conversion from a decimal fraction of a foot to inches and fractions before the physical distance can be determined with a conventional tape measure. Table 3 may be used for this conversion, showing inches and fractions for increments of 0.01 foot. For example, if a calculation yields a result of 11.63 feet, Table 3 indicates the equivalent distance is 11 feet 7-9/16 inches.

Similarly, Table 4 may be used to make the conversion from inches and fractions to decimal fractions of a foot. This table is convenient for using measured distances in equations.

Table 3
Conversion, Decimal Feet to Inches (Nearest 16th)

	Decimal Increments									
	0.00	0.01	0.02	0.03	0.04	0.05	0.06	0.07	0.08	0.09
0.0	0-0	0-1/8	0-1/4	0-3/8	0-1/2	0-5/8	0-3/4	0-13/16	0-15/16	1-1/16
0.1	1-3/16	1-5/16	1-7/16	1-9/16	1-11/16	1-13/16	1-15/16	2-1/16	2-3/16	2-1/4
0.2	2-3/8	2-1/2	2-5/8	2-3/4	2-7/8	3-0	3-1/8	3-1/4	3-3/8	3-1/2
0.3	3-5/8	3-3/4	3-13/16	3-15/16	4-1/16	4-3/16	4-5/16	4-7/16	4-9/16	4-11/16
0.4	4-13/16	4-15/16	5-1/16	5-3/16	5-1/4	5-3/8	5-1/2	5-5/8	5-3/4	5-7/8
0.5	6-0	6-1/8	6-1/4	6-3/8	6-1/2	6-5/8	6-3/4	6-13/16	6-15/16	7-1/16
0.6	7-3/16	7-5/16	7-7/16	7-9/16	7-11/16	7-13/16	7-15/16	8-1/16	8-3/16	8-1/4
0.7	8-3/8	8-1/2	8-5/8	8-3/4	8-7/8	9-0	9-1/8	9-1/4	9-3/8	9-1/2
0.8	9-5/8	9-3/4	9-13/16	9-15/16	10-1/16	10-3/16	10-5/16	10-7/16	10-9/16	10-11/16
0.9	10-13/16	10-15/16	11-1/16	11-3/16	11-1/4	11-3/8	11-1/2	11-5/8	11-3/4	11-7/8

Table 4
Conversion, Inches and Fractions to Decimal Feet

	Fractional Increments							
	0	1/8	1/4	3/8	1/2	5/8	3/4	7/8
0-	0.000	0.010	0.021	0.031	0.042	0.052	0.063	0.073
1-	0.083	0.094	0.104	0.115	0.125	0.135	0.146	0.156
2-	0.167	0.177	0.188	0.198	0.208	0.219	0.229	0.240
3-	0.250	0.260	0.271	0.281	0.292	0.302	0.313	0.323
4-	0.333	0.344	0.354	0.365	0.375	0.385	0.396	0.406
5-	0.417	0.427	0.438	0.448	0.458	0.469	0.479	0.490
6-	0.500	0.510	0.521	0.531	0.542	0.552	0.563	0.573
7-	0.583	0.594	0.604	0.615	0.625	0.635	0.646	0.656
8-	0.667	0.677	0.688	0.698	0.708	0.719	0.729	0.740
9-	0.750	0.760	0.771	0.781	0.792	0.802	0.813	0.823
10-	0.833	0.844	0.854	0.865	0.875	0.885	0.896	0.906
11-	0.917	0.927	0.938	0.948	0.958	0.969	0.979	0.990

FEEDBACK

Please use this form to give us your comments on this book and what you'd like to see in future editions.

Where did you purchase this book?   □ From ARRL directly   □ From an ARRL dealer

Is there a dealer who carries
ARRL publications within: □ 5 miles   □ 15 miles   □ 30 miles of your location?   □ Not sure.

License class:

□ Novice   □ Technician   □ Technician with HF privileges   □ General   □ Advanced   □ Extra

Name

_____ Call sign_____

Address _____

City, State/Province, ZIP/Postal Code _____

Daytime Phone (   ) _____   Age _____

If licensed, how long? _____   ARRL member? □ Yes   □ No

Other hobbies _____

_____

Occupation _____

From _____

_____

_____

EDITOR, W1FB's ANTENNA NOTEBOOK
AMERICAN RADIO RELAY LEAGUE
225 MAIN ST
NEWINGTON CT 06111-1494

please fold and tape